guide to home air conditioners and refrigeration equipment

BERNARD LAMERE

HAYDEN BOOK COMPANY, INC.
Rochelle Park, New Jersey

Library of Congress Catalog Card Number 63-17376

Printed in the United States of America

25 26 27 28 29 PRINTING

80 81 82 83 84 YEAR

Preface

The large scale utilization of equipment which employs mechanical cooling is a phenomenon of our time. While the air conditioner and refrigerator are the devices most readily brought to mind by the home owner, countless other machines which use the same basic principles are found in commercial service and in industry. We all benefit in many ways from the applications of mechanical cooling; indeed, modern industrial civilization as we know it could not function otherwise. It invades nearly all fields of human endeavor. In the processing and preservation of foodstuffs, as well as the manufacture of textiles, books, drugs, and a host of other products, even use in space craft, mechanical cooling has become very much a part of the picture.

It is felt that a need exists for a text that will explain in clear, easy-to-understand terms, the way in which a machine can use electric power to produce a cooling effect. After describing the basic principles, the book suggests simple remedies for many of the more common complaints that might be encountered. This will enable the homeowner to save himself considerable inconvenience in many situations, not to mention the matter of possibly saving a considerable loss of money by spoilage in the event of a freezer breakdown. In the field of home air conditioning, the various types of equipment are described and some space is devoted to how a particular type of equipment may be selected for use according to the construction of the building involved.

The book goes on to explain how the more complicated repairs may be made, for the benefit of those whose technical interest calls for a more complete understanding. Last but not least, it tells how the so-called gas refrigerator, which uses heat to produce cold, operates.

Bernard Lamere

Ludlow, Vermont
May, 1963

Contents

1—How Cooling Equipment Works

The value of ice for the preservation of food has been recognized for hundreds of years. In these days, when ice cubes can be made in the common household refrigerator, it is hard to imagine that, years ago, ice was so sought after that the exportation of natural ice from New England ports became a major item of commerce. Around 1810 Cuba was paying the top rate, $500 per ton!

Yankee ship owners hired hard-bitten sea captains who, in consideration of the fantastic profits, drove their vessels mercilessly to and from the ice fields. When it is realized that Boston, one of a number of ports in the business, shipped 65,000 tons of natural ice in one year, it is apparent that a large number of people were engaged in the cutting, handling, and transportation of this now-common commodity.

But, even as the New England ice shippers built their mansions, the seeds of their financial undoing had been sowed. The beginning of it was in the thinking of some European physicists in the years before the American Revolution.

They knew that when a kettle of water was placed over a fire, the heat would make the water boil and it would all eventually turn to vapor, disappearing in the air. And, they knew that if they held a cold piece of sheet metal over the rising vapor, drops of water would accumulate on the metal (see Fig. 1-1).

On the theory that all liquids act in the same way, they stated a law of physics: heat is required to make a liquid boil away (vaporize), and cooling will cause the vapor to return to its original liquid state (condense). They found this was generally true, though some liquids, such as ammonia and alcohol, would vaporize at an exceedingly low temperature. They gathered that if a kettle of alcohol would vaporize *without*

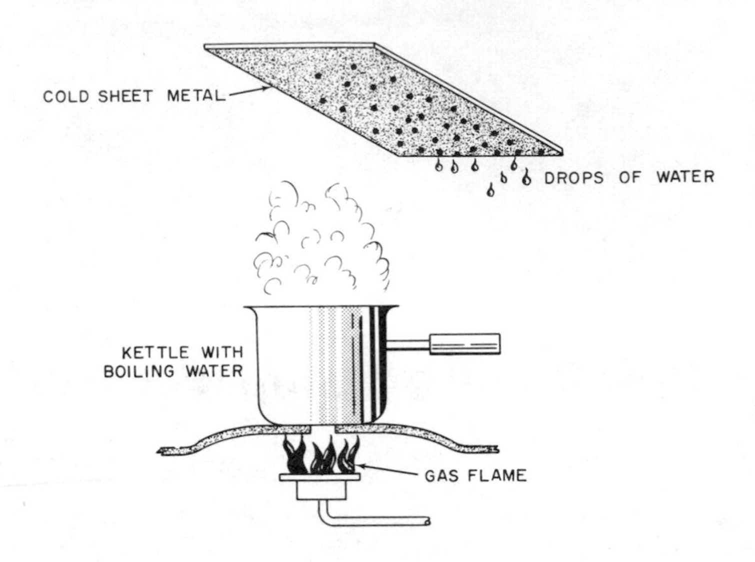

Fig. 1-1. Water vapor (steam) will condense on a cold object.

being put over a fire, it must be *absorbing* heat from the surrounding air (see Fig. 1-2).

From other experiments they learned that the vapor occupies a considerably larger space than the original liquid. Or, to state the reverse of it: the volume of a vapor is reduced when it is cooled to a liquid. They further theorized that if the vapor could be mechanically "squeezed," it would give off the heat that it must have absorbed.

They concluded that if they could take a liquid with a very low boiling (or vaporizing) point and put it in a container so fitted that the vapor would feed into a mechanical "squeezing device," pushing the somewhat compressed and much warmer vapor into a sort of radiator where it could give off its heat, the vapor should return to a liquid condition. Then, by arranging for the liquid to flow back into the vaporization chamber and repeat the cycle continuously, one could utilize the cold area of the vaporization chamber for cooling purposes.

They built a device using this theory, using ammonia (which will vaporize at a temperature below the freezing point of water) as the liquid or "refrigerant." It worked—the machine proved the theory to be correct. This principle of operation has been used in mechanical cooling equipment ever since.

To say that the first ice machines were not an unqualified success is an understatement. The ammonia had a corrosive effect on the cast iron parts, resulting in all sorts of malfunctions. After a while many bronze and brass parts were used because they would not "freeze" together the way iron did. By 1855, equipment had been devised that performed nicely, finding extensive use in breweries and meat-packing plants.

On this side of the ocean, a Mr. Twining was trying to build and sell ice machines. He did not have much success because, for some reason, people had the misconception that artificial ice was unhealthy. He did open a plant in Ohio, but he found general opposition to machine-made ice throughout the Northeast.

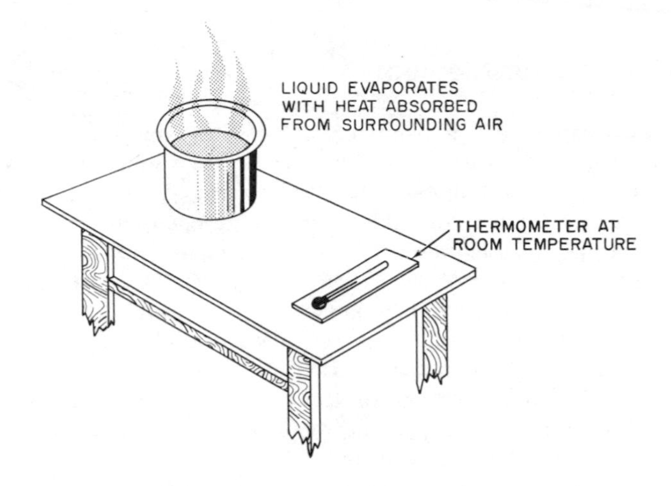
LIQUID EVAPORATES
WITH HEAT ABSORBED
FROM SURROUNDING AIR
THERMOMETER AT
ROOM TEMPERATURE

Fig. 1-2. Alcohol will vaporize from the heat of the surrounding air.

Before the Civil War, the Southerners had depended on natural ice from New England. With the advent of hostilities, the supply was shut off by the blockade. As the war progressed, more soldiers were dying of disease than bullets, and ice was desperately needed to treat fever sufferers, particularly those stricken with malaria.

It happened, as an accident of history, that one of the two countries that did not observe the Neutrality Law, made the ice machine available to the Southerners. The Rebels were faced with a difficult choice of whether to buy ice machines or gunpowder. They decided to purchase some ice machines. And so the firm of Carre, Ltee., in Paris, had some of its crates labeled: "For export to the Confederate States of America."

The invading Union Army found artificial ice being used in hospitals, and they learned that it did not make the patients worse than they were before. When word of this spread to the North, it helped to break down the widespread prejudice against machine-made ice.

In the decades after the war, there were occasional winters in New England that were warmer than usual, with the result that natural ice was in short supply the following summers. Doctors began to advise that it was safe to use artificial ice. By the 1890's, a considerable number of "ammonia plants" were in operation.

These ammonia machines were powered by steam because that was the most reliable source of power available at that time. Because a steam plant has to be constantly attended, the ice had to be produced on a large scale to be a financial success.

refrigerators for the home

It was apparent to the manufacturers of these machines that a small scale version of the ice machine for food-preservation purposes in the home would be very desirable. The development of reliable, self-starting, electric motors brought this dream closer to reality because the early manual-starting motors were of little value for home refrigeration.

The next step was to decide on the physical arrangement of the parts inside the cabinet. They knew the housewife did not want to freeze all her food; most of it merely needed chilling to 40° F. or so. Therefore, it followed that the freezing should take place in a relatively small box (for freezing water or ice cream), and the box could chill the air in the remaining part of the cabinet. And, because cold air settles, the "freeze box" had to be at the top.

The big, commercial ice machines used ammonia as the substance, or "refrigerant," that was pumped around through the system of tubing to produce a freezing temperature at the desired point. The ammonia worked very well, but it was apparent that it was not practical to use in a small household model. After the ammonia turns to vapor, it is very difficult to make it return to the liquid state. So difficult, in fact, that the equipment for mechanically "squeezing," or compressing, the vapor had to be so massive that, if used in a home, it would occupy the whole kitchen.

At that point, a technological "breakthrough" was achieved. Chemists had known for a long time that a gas called sulphur dioxide could be used in place of ammonia. It readily becomes a liquid when "squeezed," or compressed, and it gives off heat in the process. Then, as a liquid, it evaporates (vaporizes) to a gas, absorbing heat in the process. In other words, it functions as a refrigerant. They thought that, if "refrigerant" meant "a substance used to produce cold," the entire unit should be termed a "refrigerator." The first household refrigerators were marketed in the years after World War I, and, though they have changed in appearance and are vastly improved in efficiency, the basic principle of operation remains the same.

cooling system components

People proceeded to assign names to the refrigerator components. The box-shaped container in which the refrigerant vaporized (or evaporated) was, technically, an evaporator, but almost everyone called it a freezer (a term still used today). From the freezer, the then heat-laden vapor went to a device which, by "squeezing" reduced the vapor in volume. This "compression," as it is termed, took place in a cylinder that was closed at one end with a piston moving up and down in it. A crankshaft and connecting rod moves the piston (see Fig. 1-3). The

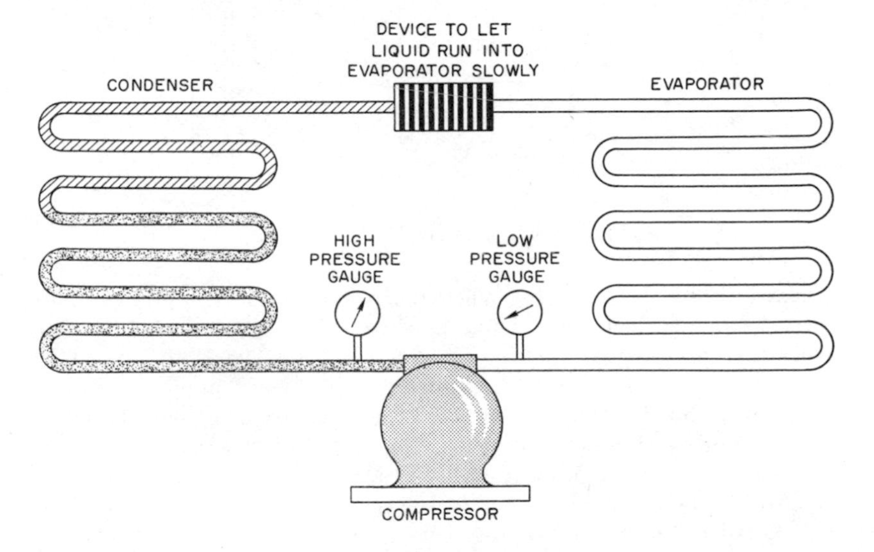

Fig. 1-3. The refrigerant cycle.

crankshaft had a pulley on one end, connected to an electric motor by a belt. This "up-and-down" movement, or "reciprocating action," led to naming the device a "reciprocating compressor."

The flow of refrigerant in and out of the cylinder was governed by "one-way valves." In Fig. 1-4 we see how the valves, which are flat

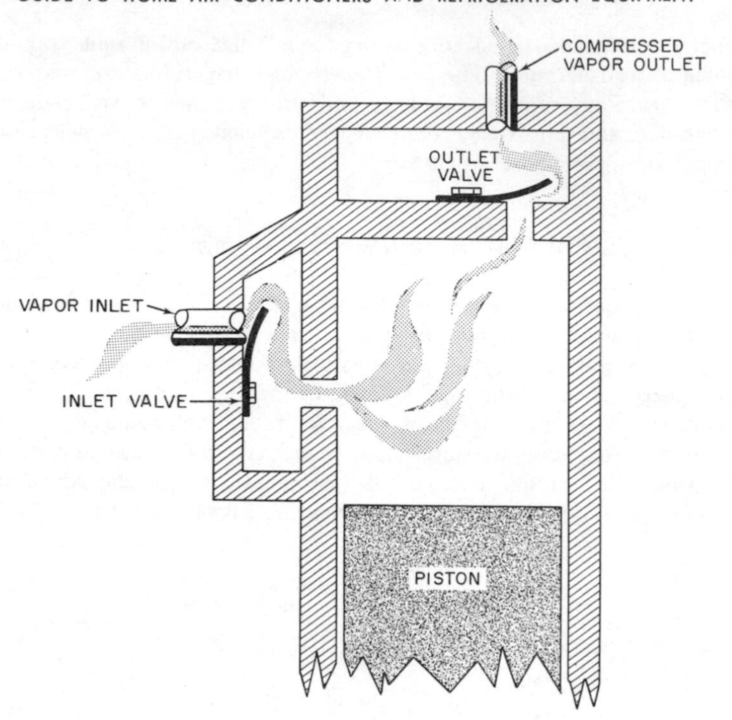

Fig. 1-4. Operation of the "one-way valves."

pieces of spring steel, work. When the piston in the compressor shaft goes down, the inlet valve bends to allow vapor to be drawn in, and, when the piston goes up, the inlet valve bends back to cover the inlet. The vapor then will be pushed out at higher pressure through the outlet, and the outlet valve will then cover the opening to prevent the vapor from returning through it. The cycle is repeated.

Because compression made the vapor hot, and this heat must be removed in order for the vapor to return to the liquid state, a method had to be devised to remove this heat of compression. The vapor flowed through some tubing, which in turn had air flowing over it. This air cooled the hot vapors. Inside the tubing, the cooled vapor formed drops of liquid that trickle to the bottom (see Fig. 1-5). The compressor pushes the vapor into the tubing so fast that a certain amount of pressure is maintained. Enough, in fact, to cause the liquid to run into the freezer to repeat its cycle. This process of a vapor changing to a liquid because of cooling is called "condensation," and the part of a refrigerator where it takes place is called a "condenser."

In the old refrigerators, the freezer, compressor and condenser, and the tubes connecting them, were put together with bolts and screw-on fittings. This arrangement is called an "open" system, as distinguished from a modern "sealed" system, where the connections are soldered, or "sealed," so that the refrigerant cannot get out and air cannot get in.

We will not devote much space to the study of open reciprocating mechanisms because, at least for household use, they are rapidly disappearing from the American scene. Sulphur dioxide is corrosive and smells horribly (like rotten eggs) when it escapes. It served its purpose at the time, however.

In the so-called open units, the flow of refrigerant is governed by a float valve. One type of float-valve assembly consists of a tank with two openings. The larger opening is a pipe through which liquid refrigerant moves freely (see Fig. 1-6). Inside the tank there is a float that rises and falls as the liquid rises and falls. When the float drops, the leverage action of an arm attached to it causes a needle point to move away from its valve seat, opening a small hole through which liquid could flow into the tank. As the float rises, the procedure is reversed with the reseating of the needle point, preventing more liquid from running in until called for by another drop in the liquid level.

All float valves can develop trouble. If the float leaks and fills with liquid, the needle point will be drawn to the end of its travel. Or, the point might adhere to its seat and not open at all. If a particle of metallic dust lodges between the needle point and the seat, it will not

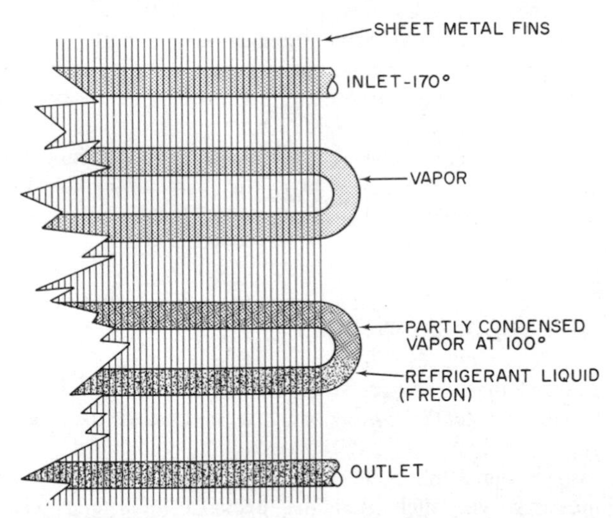

Fig. 1-5. **The old-style condenser with bottom receiver tank illustrates how refrigerant vapor condenses to a liquid when cooled.**

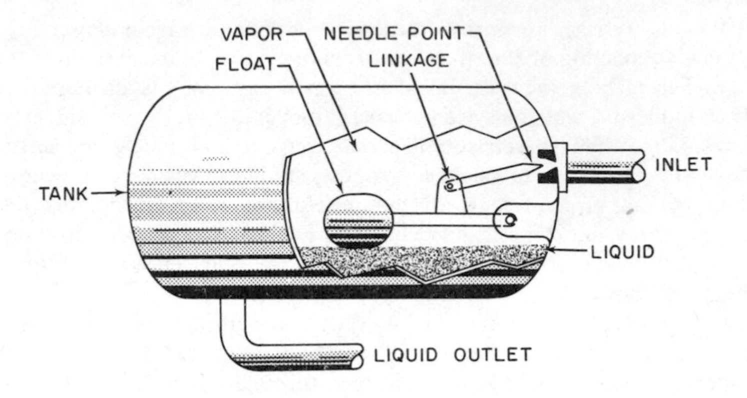

Fig. 1-6. The float valve.

close properly. Needless to say, these controls are a source of considerable trouble.

One of the early refrigerators was called a "low side" system. The freezer had a tank (with float valve) arranged so that the liquid could run into the sheet metal (or perhaps cast metal) passages of the freezer. The term "low side" was coined because the pressure of the vapor in the freezer was low compared to the pressure of the entering liquid. The needle point may be thought of as the dividing line between the low and high pressure parts of the system. This idea is worked out in all forms of refrigerating equipment; its present day application will be discussed later.

Thus, we may visualize the refrigerant as running into the freezer where it evaporates, absorbing heat from the surrounding air (through the freezer walls), and then being drawn (as heat-laden vapor) to the compressor.

why the hot vapor condenses

We mentioned that a vapor will condense to a liquid if it is cooled; that is, it will *eventually* condense if it is cooled to a low *enough* temperature.

In the case of the sulphur-dioxide refrigerant, if the amount of vapor in the condenser was such that the pressure of the vapor was the same as that of the outside air, the condenser would have to cool the vapor to at least 14°F. to make it become a liquid. Because that temperature is way below room temperature, we might ask, "Why *does* it go back to the liquid state?"

The answer to this question is found in a previous statement that "a certain amount of pressure is maintained." It is a principle of physics that the more you compress a vapor, that is to say, the more you raise its pressure, the higher you raise the temperature at which the vapor will liquify or condense. In the case of sulphur dioxide, by having the vapor at a pressure of 60 pounds per square inch (psi) or more, enables it to condense at 75 or even 90°F.

The reason that the pressure in the condenser can build up so high is that the float valve holds back the refrigerant and lets it run into the freezer very slowly. Thus, the float valve serves two functions: on the one hand it lets the liquid go into the freezer only as fast as it can evaporate, and on the other it allows pressure to build up in the condenser.

Another early type was the "high side" system. The float was in a "receiver" tank underneath the condenser. In this case, the float let liquid *out* when the float was raised. The control permitted just enough liquid refrigerant to keep the pipe or tube to the freezer (the "liquid line") and the freezer itself filled. With this arrangement, small bubbles of vapor could form on the freezer walls and rise to the top to be drawn off. Whether this happens as shown in Fig. 1-7A, or whether a drop of liquid runs onto a flat surface and completely evaporates, as shown in Fig. 1-7B, the result is the same: heat is absorbed.

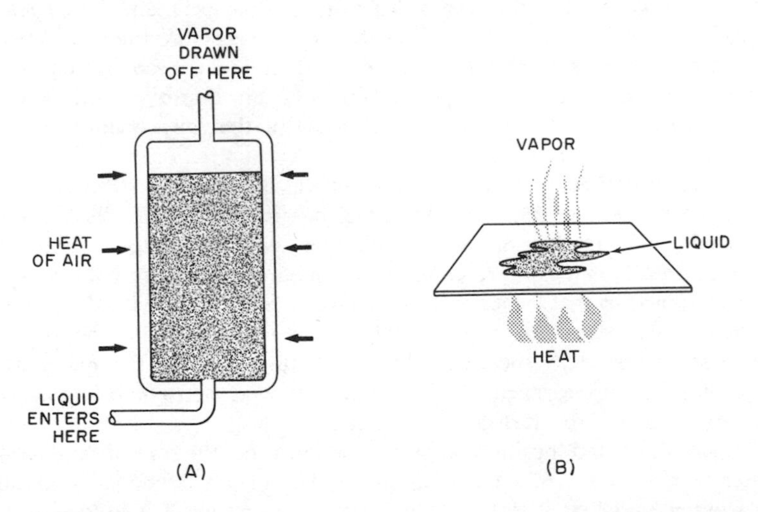

Fig. 1-7. An evaporating liquid absorbs heat.

The float valves mentioned above were intended only to regulate the flow of refrigerant through the system; how long the compressor ran was another matter. That was controlled by a bellows switch on the "low side." When the low-side pressure changed to a predetermined extent, the bellows would contract (or expand), moving a toggle arrangement that would open (or close) the contacts of the power supply to the motor. This method did maintain a fairly constant temperature in the cabinet, but, if a different temperature was wanted, it was necessary to remove the panel at the bottom of the refrigerator and use tools to make the adjustment—hardly a convenient arrangement.

But the main disadvantage of the sulphur dioxide equipment was that, on the low side, the vapor pressure would be *very* low; so low, in fact, that it was actually less than that of the atmosphere. This "negative pressure," or "vacuum," as it is termed, would draw in air any place where there was a leak. The open systems had a multitude of fittings and connections; any of them could leak. Leaks on the high side would simply cause loss of refrigerant; on the low side they would as we said, cause air to be drawn in and mix with the refrigerant. And the metallic dust that wore off of the bearings and piston could cause more trouble.

the sealed unit

It was apparent that the way to end the leakage problem (or at least reduce leakage to a reasonable point), would be to enclose the mechanism in a leak-proof shell with the freezer and condenser connected with soldered joints. Because the sealed unit was an improvement, it was adopted throughout the industry. It is doubtful that any open-type unit was made after 1935.

The General Electric Company devised the Monitor Top Refrigerator and marketed it in 1925. It featured a simplified compressor with hardened parts. The condenser was circular with the motor-compressor assembly inside of it and the controls mounted outside. The cold control was redesigned so that it had only an electrical connection with the motor.

These parts rested on a panel, which, actually, was the top of the refrigerator cabinet. Suspended from the bottom side of the panel was the freezer. In appearance, the Monitor Top was conspicuous because it extended above the refrigerator cabinet.

The top-mounted position allowed the heat of the motor and condenser to rise away from the food space. The idea is technically sound, but women said that it did not "look nice", and wanted it hidden from view. It was moved to the bottom, which, of course, is where all companies put it today.

"Freon," the modern refrigerant

In the sulphur-dioxide era, much research was done to devise a substiute for the foul-smelling gas. When methyl chloride became available, it was viewed as a blessing. It (normally) maintained a positive pressure in all parts of the system and did not smell badly when it leaked. Its popularity was brief, however, as it lost favor when the DuPont chemists invented the "Freons", which nearly all household units use today. We shall refer to modern refrigerant as Freon, although the Allied Chemicals and Dye Company (and some other concerns) are making a similar product with a different name.

There are several varieties of Freon, differing in their vaporizing temperature. Freon-22, for instance, boils at about 41° below zero. It is difficult to think of anything *boiling* at such a low temperature, however, it is not uncommon for most gases. Freon-113 boils at 117° above zero and Freon-11 at 75° above. From this, we can conclude that in a room at 76° F. it would be easy to liquify Freon-11, and that in a room at 74° F. it would be easy to boil it. It is a very complex matter, however, to apply these ideas to the actual design of a refrigerator and to say precisely what degree of cold will be obtained with a given set of conditions. The internal pressures determine performance, and you will have a different pressure at every inch of the circuit; all the bends and restrictions will change the pressure. Furthermore, you will have the radiated heat of the motor to contend with, as well as the circulation of air, which might be either warm or cool. As a matter of fact, the computations are so involved that rapid refrigeration research had to await the development of the electronic computer.

For our purposes, it will suffice to say that a unit is supposed to have a particular type of Freon in it, and, if a different type is used, the unit will perform poorly or not at all.

modern controls

With the advent of the sealed unit, a change was made in the starter switch. The belt-connected motor had a centrifugal switch that disconnected the starting winding when the motor speeded up enough to move a weight away from the shaft. Obviously, in a welded motor-compressor housing you would not be able to get at it for repair. The present switch, mounted externally for this reason, is called a "relay." In addition, the troublesome float valve was replaced with a "restrictor," of which we will learn more later. The operation of the sealed unit is regulated by a thermostatic switch, or "cold control."

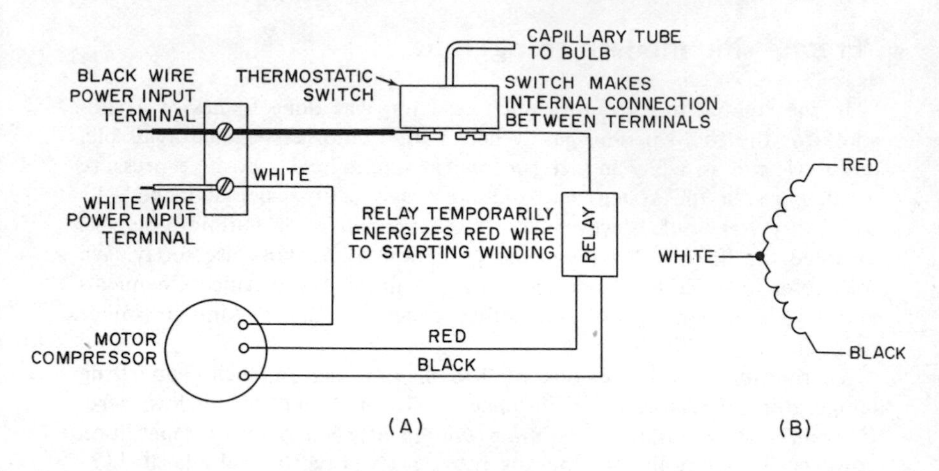

Fig. 1-8. A 115-volt basic circuit.

The sealed unit, relay, and cold control are the basic working parts of the conventional refrigerator (see Fig. 1-8). Engineers have been able to take these "basic parts" and somewhat modify them and make new devices for other purposes. One of the most important of these is the air conditioner.

2—A Look At Air Conditioning

Keeping cool in the summer must have been one of man's earliest endeavors. The Babylonians had their own version of how to reduce the temperature in a home by running cool water in stone conduits under the floors. (They also warmed the water for radiant heat in the winter!). From then until Thomas Edison died while telling his listeners how to air condition the room he was in, no one thought of a practical device to combat the summer heat.

The present-day home equipment is a byproduct of temperature and humidity control, and air filtering developed for industry. In the beginning, comfort was only incidental. But, it would be quite proper to say that the modern home air conditioner is a "comfort-making machine." It has several functions that contribute to this end. The first, obviously, is cooling the air, which conserves body energy by lightening the load on the circulatory system.

The second function is dehumidifying the air. We mentioned that warm air can absorb and hold more moisture than cool air. But, without knowing the statistics, it is difficult to realize how very sharply the moisture holding capacity of the air increases as the temperature goes up. A pound of water weighs 7,000 grains. At 70° F. a pound of air will hold 110 grains of water. But the figure jumps to 156 at 80° F., 218 at 90° F., and 302 grains at 100° F. Thus, with 100% humidity you would have almost three times as much moisture in the air at 100 as at 70° F. Perspiration cannot evaporate as readily when the air is already moisture laden, and this produces the well-known "sticky" and uncomfortable feeling.

Another fact of nature is that rainstorms occur when a cold front collides with a moisture-laden air mass—simply because the water vapor condenses to rain when the air is cooled. Applying this principle to air conditioning: when the moist hot air passes through the cold evaporator, some of its moisture is deposited.

The third function of the device is to filter the air. This makes the house easier to keep clean, of course, and it makes for healthier living conditions by removing, or at least reducing, air-borne pollen and irritating industrial dust. This has proven a boon to persons suffering from respiratory diseases.

A fourth function is to circulate the air, breaking up stagnant "pockets" and bringing about an even temperature in the various parts of the room.

Some units perform a fifth function—providing heat when the temperature drops below a specified point.

types of air conditioners

There are several types of home air-conditioning units. The earliest was the window type, which was readily adapted for "through-the-wall" installation. There also is the "free-standing" type, which stands away from the walls and windows and takes in, conditions, and expels the air without benefit of any ductwork for either the air supply or distribution. These are used mostly in places of business that have large rooms to take advantage of the relatively large capacity of these units.

The latest development, and one that is finding wide acceptance throughout the country, is "central" air conditioning, in which cooling coils are placed in forced-air ducts so that the cool air can be distributed throughout the house. There are several methods of installation. The location of the cooling unit itself, or parts of it, and the size and placement of the ducts is influenced by the design and structure of the building. It is a matter of determining what installation is compatible with economical operation of a system that will provide efficient cooling. This is a complex matter that has been given considerable study.

In new construction, several factors are given more consideration than they were in the pre-air conditioning days. To quote Carrier's "Home Management" booklet:

> "Principles relating to utilization and control of heat from the sun become extremely important in an air conditioned home. Careful attention should be given to exterior design, orientation, size and placement of windows, colors, insulation, and proper selection and installation of air conditioning equipment. By following this approach, one builder saved one-third in first cost of air conditioning and one-half of annual operating cost in a house of medium price."

We have said that the basic principles of operation of the "comfort machine" are the same as that of the refrigerator described earlier. The principle parts of the air conditioner, however, are not necessarily all in the one cabinet as they are in a refrigerator. In Fig. 2-1 we see the compressor and condenser outdoors and the refrigerant lines extending through the wall to reach the evaporator, or cooling coil, which is in the duct system of a furnace. The warm air is drawn into the unit, over the cooling coils, and distributed over the ductwork.

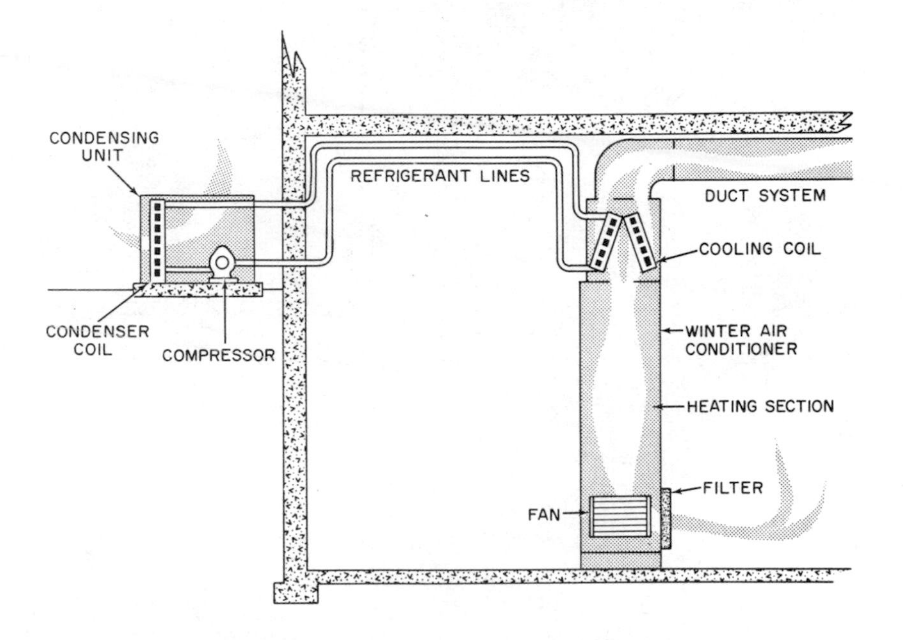

Fig. 2-1. Furnace with separate condensing unit. (Courtesy of Carrier Air Conditioning Company.)

In Fig. 2-2 we have a self-contained unit outdoors on a "pad" and connected by a collar to the sheetmetal box on the top of the furnace. Here the return air is drawn through the furnace filter and up into the cooling unit, where it is reversed in direction and pushed out into the ducts. The flow of air is governed, of course, by baffles that are not shown.

The cooling unit can be either electric or gas fired. In Fig. 2-3 we have gas used for both heating and cooling. Cooling with heat may seem inconsistent, but Chapter 9 explains how it is done. In this case, the cooling unit chills water, which is forced through pipes, cooling the duct coils as it circulates.

In Fig. 2-4 we have another system which uses water, but, in this

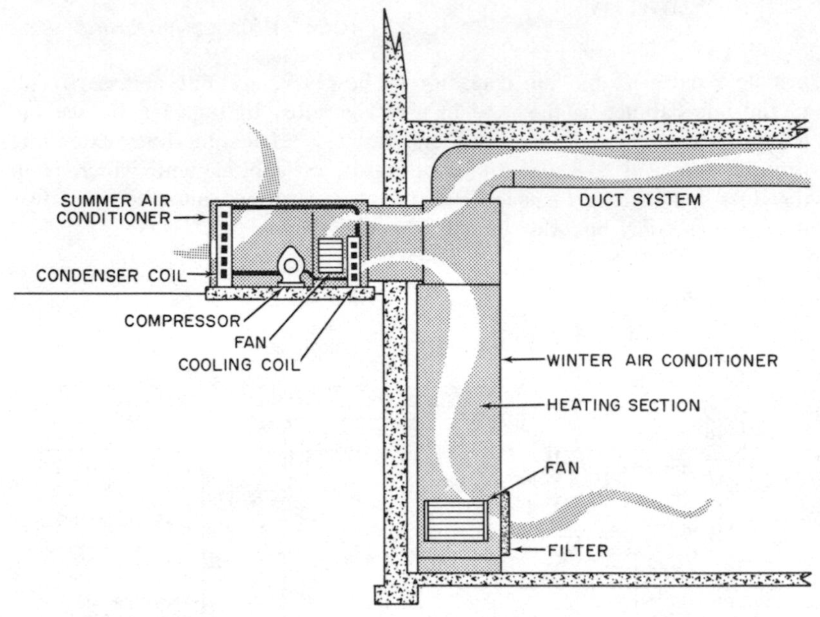

Fig. 2-2. Furnace with self-contained air conditioner. (Courtesy of Carrier Air Conditioning Company.)

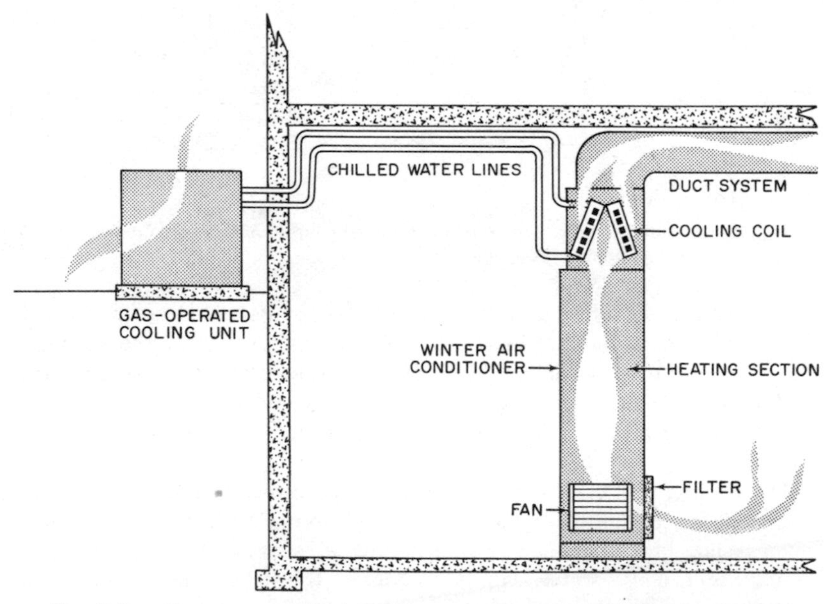

Fig. 2-3. Two-piece gas-fired cooling unit. (Courtesy of Carrier Air Conditioning Company.)

case, the water cools the condenser instead of being cooled by the evaporator. In the process, the water picks up heat, which it then must lose in a "cooling tower." The one shown rests on the ground, and uses air to remove the heat from the water. Of course, circumstances might dictate that it be placed on the roof.

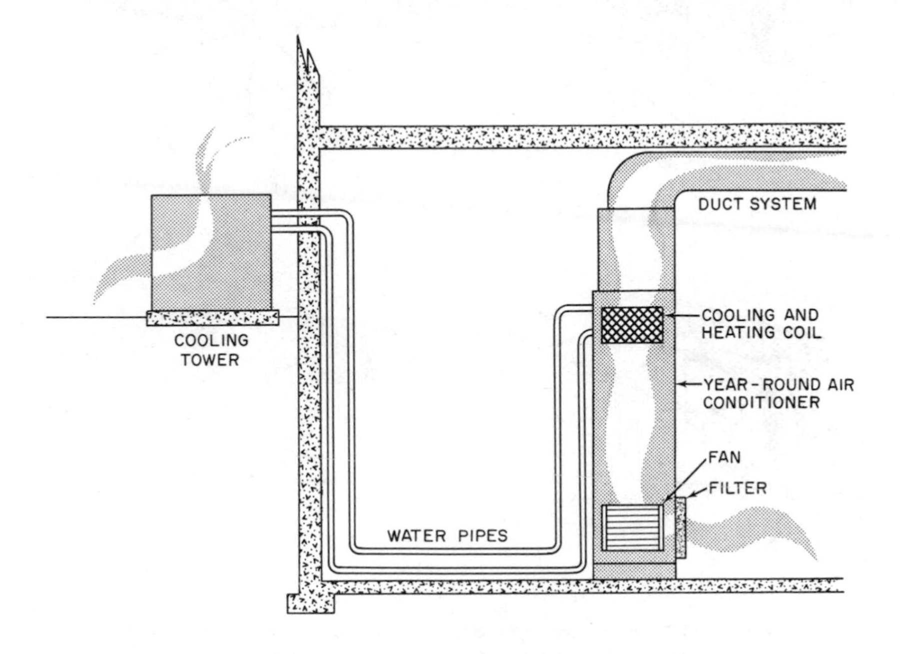

Fig. 2-4. One-piece gas-fired cooling unit.

Fig. 2-5 shows methods of air distribution. We have, in Fig. 2-5A, the unit in the basement with ducts running under the floors. Figure 2-5B shows the unit in the center of the house with ducts running through the cement-floor slab.

A perimeter system puts the cool air where it can diffuse the heat of the outside walls, thereby creating an even temperature. In Fig. 2-5C we have a center duct running *under* the ceiling. Figure 2-5D shows a method for installing a *second floor* cooling system with little alteration of the existing structure. The distribution ducts extend sideways from the center, or "trunk" duct, and direct the air downward through ceiling openings. In all of these systems, return registers and ducts are needed but are not shown.

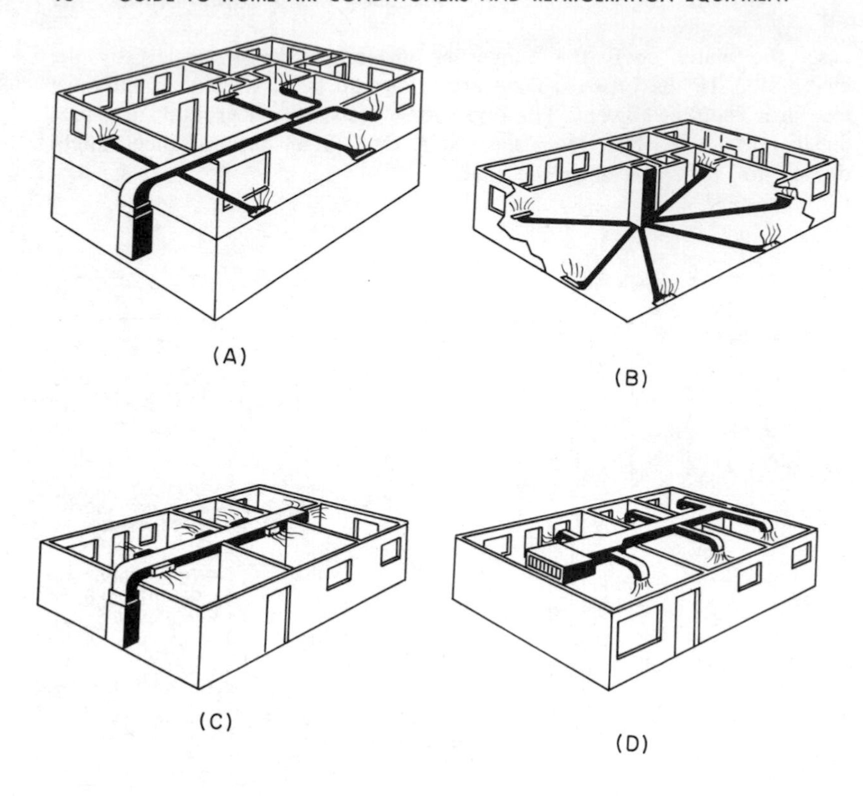

Fig. 2-5. (A) Perimeter system in basement of house; (B) perimeter system in house built on foundation; (C) center duct system; (D) attic system with ceiling registers. (Courtesy of Carrier Air Conditioning Company.)

perimeter vs. central duct system

Whether a perimeter system is used will be dictated by the severeness of the winters. With extreme cold, it is desirable to counteract the chill of the outside walls, making a perimeter system almost mandatory. If the house has a forced-hot-air system, the most economical method of installing central cooling is to utilize the supply and return ducts that are provided.

In any case, the air circulation in the conditioned spaces is governed by the shape, size, and location of the registers, as well as the capacity of the fans and ducts. How the air circulates within a room can be determined by holding a cigarette in the air stream and observing the direction in which the smoke travels.

A combination heating and cooling plant is termed a "year-round air conditioner." Where the furnace is of the steam or hot-water type, the air-conditioning system would have to be installed separately.

The foregoing does not exhaust the possible methods of installing central air conditioning; the methods mentioned are merely the most commonly used.

the window air conditioner

We already know that refrigeration is the heart of the air conditioner. Now, what we want to learn is how refrigeration is utilized and what might go wrong if the unit does not perform properly. The easiest way is to study a small unit first. An early-type window air conditioner with a simple "on-off" switch is ideal for this purpose; other than the switch and the fans, the only other parts are the sealed unit and a starter for the motor.

Because the condenser gives off heat when the unit is in operation, the first consideration is, obviously, to expel the heat from the conditioned room. It would be quite pointless to allow it to return into the room that you were attempting to cool. Removing the heat is accomplished by using two fans—one blowing air through the cold evaporator, and the other moving air over the motors, through the hot condenser and removing it from the conditioned room. Figure 2-6 shows a typical

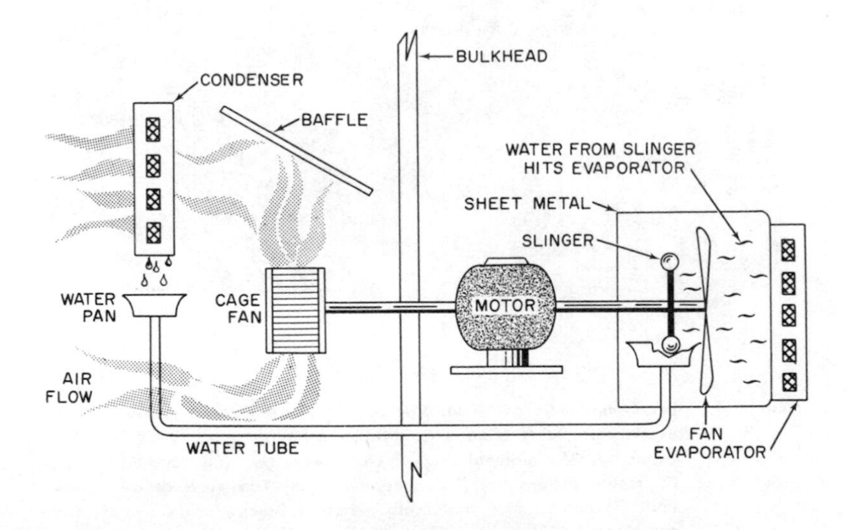

Fig. 2-6. Air flow and condensed water removal in window unit.

arrangement in such a unit, and Fig. 2-7 shows actual window units with the parts indicated.

We have now considered two conditions that produce vaporization—

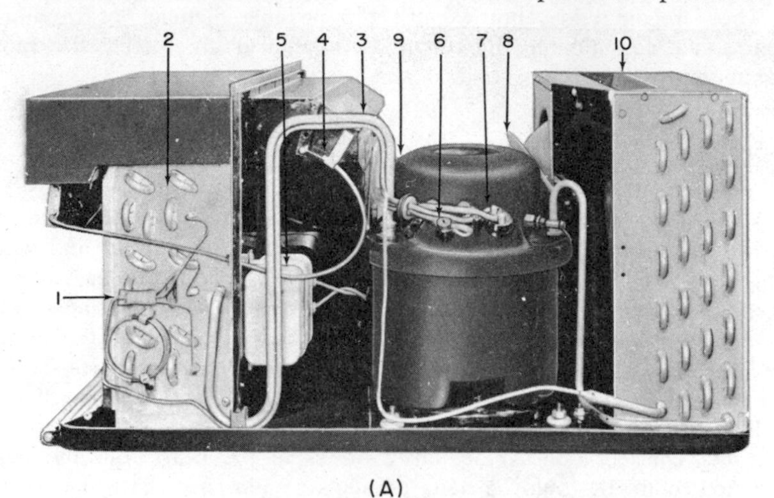

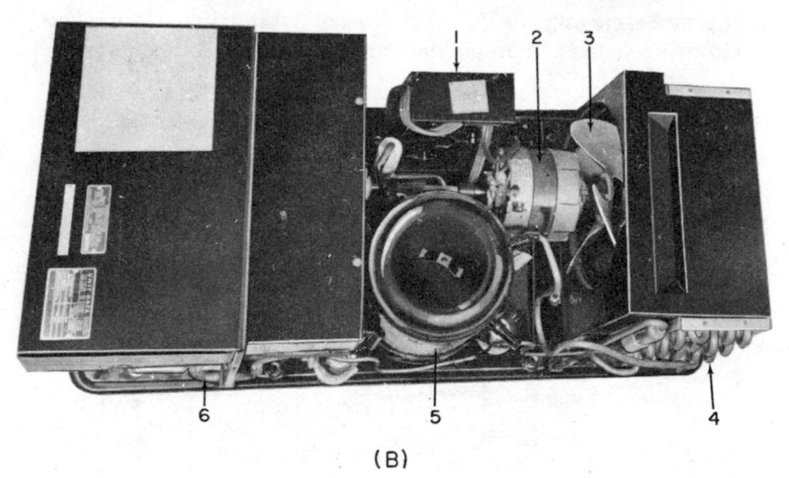

Fig. 2-7. (A) Components of a window unit: (1) liquid line distributor, (2) evaporator, (3) suction line, (4) outdoor air control, (5) running capacitor, (6) terminals, (7) motor protector, (8) condenser fan, (9) motor compressor, (10) condenser. (B) Top view of a window unit: (1) control box containing terminal blocks, switch, relay and capacitor, (2) fan motor, (3) fan, (4) condenser, (5) motor compressor, (6) cooling coil. (Courtesy of Frigidaire.)

that which takes place in the freezer compartment of a refrigerator, and that which takes place in the finned tubes of an air conditioner evaporator. A great deal of research has been devoted to the latter method of vaporization, especially in the use of a similar device to cool automobile engines. The modern finned evaporator (or condenser) is able to transfer a considerable amount of heat for its size.

Your unit might have a fan motor with a fan on each end of the shaft, or it might have a separate motor for each fan. In the latter case, the two motor casings would probably be bolted together to save space, but this would not affect their operation (see Fig. 2-8).

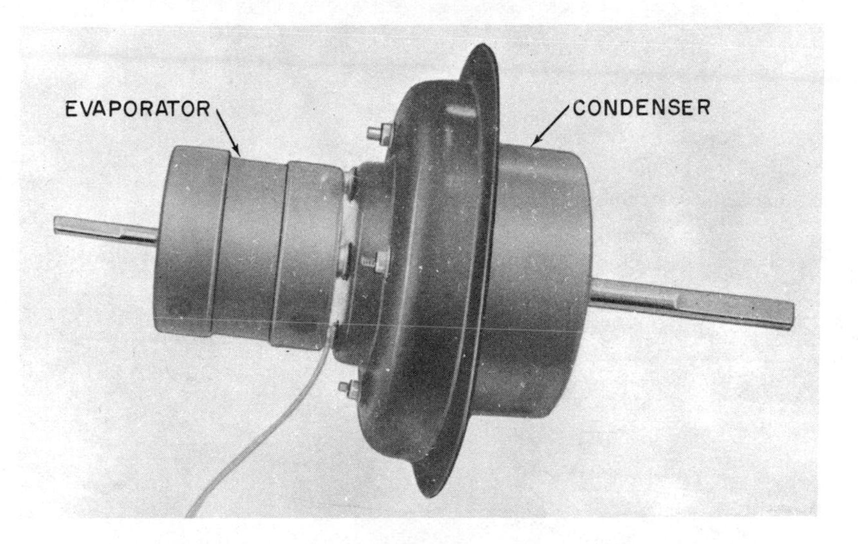

Fig. 2-8. Fan motors bolted together. (Courtesy of General Electric Company.)

The conventional arrangement has a "cage" fan for the room air. It draws in filtered air through its center opening and throws it to the periphery where it strikes a baffle and is reversed in direction so that it blows into the room after passing over the cool tubes and fins. Fresh air from outdoors may or may not be introduced to the air stream on its way to the tubes. A cage-type fan is able to develop the thrust necessary to pass the air over the tubes, through the filter, and still have it travel into the room with some force.

The condenser might have a fan similar to the one in your automobile (see Fig. 2-9).

With a large volume of air passing through the evaporator, it follows that considerable moisture will be deposited on it. On the early, small models, the moisture (or condensate) was disposed of simply by letting it drip down into some troughs that returned it to a pan under the condenser. At that point, the heat and turbulence of the air would evaporate it. If it overflowed the pan and dripped onto a lawn, no harm would be done. Dust or traces of oil might get into the water, however, so that if the wind blew it onto the building an unsightly stain might

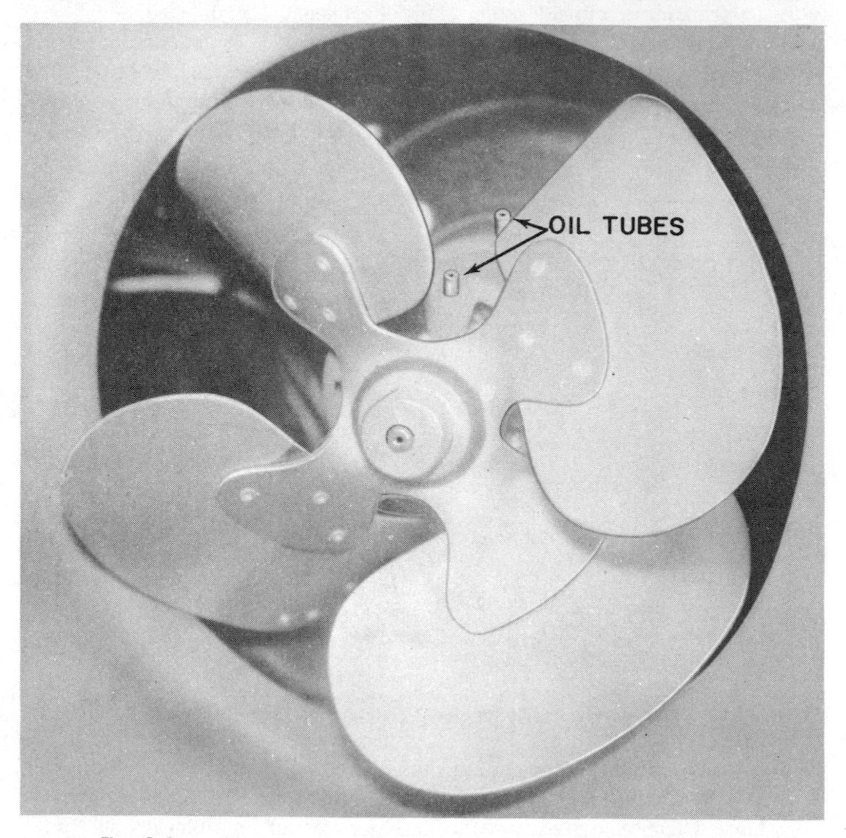

Fig. 2-9. Condenser fan. (Courtesy of General Electric Company.)

result. A more positive method of dealing with the condensate was indicated. The newer models have a "slinger ring," which is a wheel of rubber or metal, that "slings" or spatters the water onto the hot condenser (see Fig. 2-10).

Another development introduced switches for various fan speeds and various degrees of cooling. An electric heater, wired so that it would run with the compressor shut off was also added, but because the heater coils are of limited size, it should not be expected to do much except take off the chill in the spring or fall (see Fig. 2-11).

Fig. 2-10. Typical slinger system.

Fig. 2-11. Electric heater on air conditioner.

the heat pump

The next step was to develop a device so that the unit could furnish heat on chilly days without requiring red-hot heater wires or the like. It has been said that a customer wrote to a manufacturer, saying that his window air conditioner had accidently been installed backwards and he found he could use it for a heater. That is not the way it happened, but the statement has a degree of logic because the unit is, so to speak, hot at one end and cold at the other.

More than a hundred years ago an Englishman, Lord Kelvin, did the paper work on a device to take whatever heat could be found in air or water and transfer it to a warmer medium. In other words, to draw heat from air at, say, 50° F. and use it to maintain a room temperature of 70° F. A Mr. Kramer, in the modern era, was able to construct such a machine. Since then, scientists have theorized many uses for so-called heat pumps, such as drawing water from under ice and extracting heat from it and using that to heat a house. The impractical aspect of that scheme, however, is that you have to do so much pumping of water and refrigerant that the power cost would be higher than the value of the heat obtained.

But, actually, the same idea is used in your refrigerator. In the freezer compartment, the temperature is below freezing, but the unit takes heat that seeps in and expels it to the much warmer room through the condenser. It would follow that, by revolving the window air conditioner, it could be used as a heater on a cool day. It is impractical, however, to have a "merry-go-round" arrangement so the user can turn the air conditioner as desired.

Faced with this, the designers devised a way to make the condenser act as the evaporator, and vice versa. In the window air conditioner the two coil banks are similar in construction to make this substitution possible, and, with the use of a "reversing switch," the desired action was accomplished.

Today, industry is trying to eliminate the use of seals and packings on external connections. When you start putting valve stems and the like on a sealed unit, it is no longer sealed, and, with respect to leaks, it reverts to the obsolete "open" system. The problem was to devise a control for the refrigerant. This was solved by the use of a solenoid (see Fig. 2-12).

A solenoid has an electric coil, which, when energized creates a magnetic field inside the solenoid. The field extends into a nonmagnetic casing (such as brass). The force developed by the magnetic field moves a plunger inside of the casing. The plunger is spring loaded, that is, there is a coil spring inside of the casing that rests against one end of the plunger. When the coil is energized, the plunger compresses the spring, and when the current is shut off, the spring returns the plunger to its

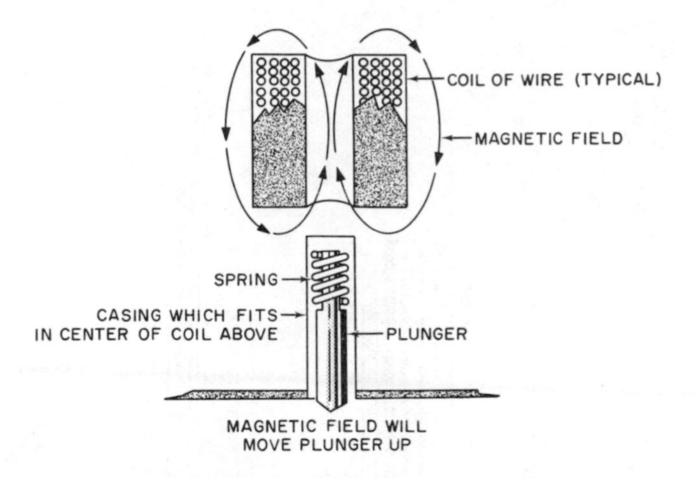

Fig. 2-12. Operation of solenoid.

original position. Energizing the coil diverts the flow of a small stream of refrigerant, which moves a piston connected to a valve.

Why not use a solenoid to move the valve directly? Well, that could be done, but, with this other arrangement, the power of the motor and the pressure of the gas (refrigerant) can be used. Consider what happens when you raise a car with a hydraulic bumper jack. Moving the handle pushes a small piston that pumps oil from its chamber into a larger chamber with a larger piston. If you pump oil into a chamber ten times as large, the piston in this chamber moves only one-tenth the distance as the smaller piston. It works like a reduction gear—you trade speed for power (see Fig. 2-13).

In using the solenoid system, the small refrigerant stream governed by the solenoid will go into, say, the left end of a cylinder which has a piston with a reversing valve (see Fig. 2-14). More likely there are two pistons with the valve, mounted on the arm between them, sliding back-and-forth as a single unit. With the solenoid letting gas into one end of the cylinder (1), the valve assembly would slide to the left, and conversely, when the gas flow was switched to the other end (2), the assembly would slide to the right.

The valve itself is simply a plate with a channel cut out. This plate slides on another plate with three holes, or "ports," that are in line and, at any time, two of them are covered by the channel in the valve plate. The center port, "B", receives the hot, highly compressed gas from the compressor. Port "A" leads to the indoor tubing and "C" to the outdoor tubing. With the valve plate slid all the way to the right (Fig. 2-14B), the hot gas would be directed into the outdoor tubing, and

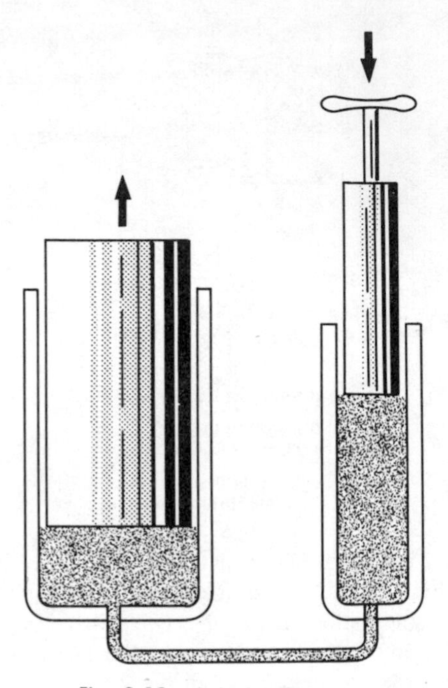

Fig. 2-13. A hydraulic pump.

similarly, with the valve plate to the left, the hot gas would go to the indoor tubing.

There is no question that this arrangement can provide warmth, but it is not a substitute for a heavy-duty heating plant.

Various companies have different notions about the automatic control of the hot-gas and electric features. Some newer models have two electric resistance heaters, and the "program" for control might go like this: With the switch on "heat," if the room temperature drops 2° below the thermostat setting, a 2200-watt heater will be energized. If it continues to drop, the compressor and reversing valve will be energized. If the outside temperature drops to 46° F., the compressor will shut off and the second heater will go on. (That requires another thermostat outdoors, wired so that it can break the circuit to the compressor.)

the defrost heater

When these units were first operated, it was found that considerable frost deposited on what was, for cooling purposes, the evaporator. That called for a defroster. If the model had a radiant room heater, that was

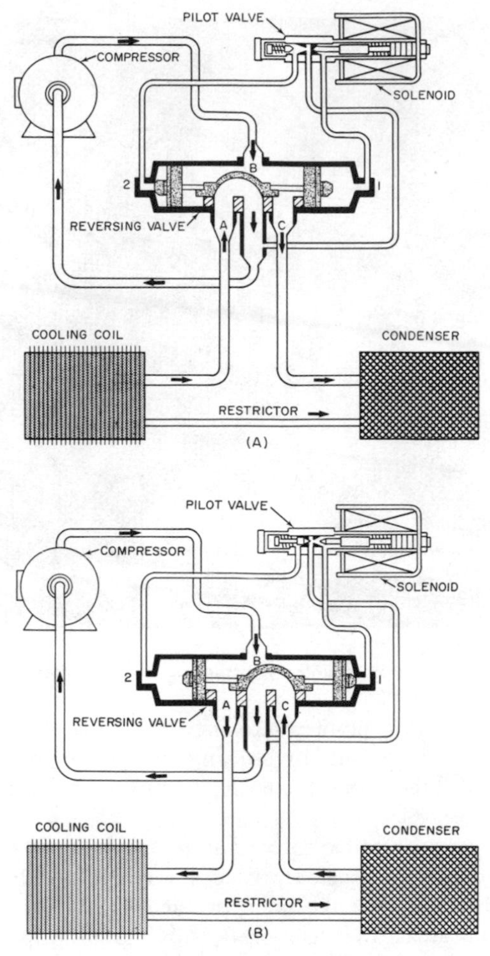

Fig. 2-14. (A) Heat-pump cooling circuit; (B) heat-pump heating circuit. (Courtesy of Frigidaire.)

utilized as the defroster. If not, defrosting was accomplished with a radiant (red-hot wire) heater of smaller size. The defroster control switch was either on the suction line (see Fig. 2-15) or on the evaporator (see Fig. 2-16).

A window location was inconvenient for some people, resulting in the manufacture of through-the-wall air conditioners. In new buildings, many contractors provide space for this type of installation.

Before you 'do" anything to a window air conditioner, carefully study the mounting plates, brackets, and fixtures. In some cases, these metal

Fig. 2-15. Defrost control mounted on suction line. (Courtesy of Frigidaire.)

parts have corroded to a point where they can "give way" if the unit is subjected to slight movement. By shifting the unit to remove panels or parts, you might break the mounting fixtures, creating a hazardous situation.

Some metals are susceptible to corrosion at any location, and in some sections of the country corrosion proceeds very rapidly. Industrial fumes contribute to corrosion, but the main problem is salt carried by the air. In certain coastal areas, the deterioration of mounting parts is so rapid that stainless-steel parts eventually might have to be used for safe mounting. Aside from the corrosion problem, some window units are not very securely installed in the first place.

improper cooling

When a window conditioner does not cool properly, there are several possible reasons *not* related to the sealed unit or the controls. It is surprising how often a repairman is summoned only to find that drapes or venetian blinds are covering the front of the unit, thereby restricting the flow of air. And there are some other causes of poor cooling that are neither mechanical nor electrical. For instance, the condenser can become covered with insects. Certain insects can cause a great deal of trouble because, in cool weather, they will look for a warm area. They can get inside the cabinet, and, when the unit starts, the fan will blow many of the insects onto the tubing and fins. Most of them can be removed with a vacuum cleaner.

Another cause of poor performance might be a filter that is covered with dust. Or maybe the filter has been accidently left out of the unit,

Fig. 2-16. Defrost control mounted on evaporator. (Courtesy of Frigidaire.)

thereby allowing the evaporator to become covered with dust. Many types of filters are in use. There is the old disposable kind, of course, and some plastic and pressed-metal types that, it is said, can be washed and re-used. There is a sponge-rubber filter that, the manufacturer claims, will "arrest the movement of pollen and bring heaven on earth to hay-fever sufferers."

The admixture of outdoor air is governed by a baffle. Although the baffle is supposely rugged, it could warp, thereby allowing hot air to come in at a time when it is not wanted. Perhaps the linkage between the baffle and the knob is damaged. Or the knob might possibly have turned on the shaft so that, in the "recirculate" position, the baffle does not close. For that matter, the fill-in panels between the unit sides and window casing might be loose, allowing hot outside air to enter. Or it might be that the unit simply is not big enough. In this case, the fact will not become apparent until a heat wave comes along.

water leaks

Water leaks are another source of concern. With excessive humidity, the water removed from the air might be considerable (several pints per hour). If the drain tubes become plugged or kinked, the water will back up and run onto the floor. In Fig. 2-17 we can see how water can run from the pan (2), through the tube (5), to the connection (6) that carries the water through the bulkhead, to another tube, and then to a pan under the slinger ring. If the tube (5) is not in place in the bulkhead fitting, the water will not be carried away. Or, if the front of

Fig. 2-17. Bulkhead in a window unit: (1) evaporator, (2) drain pan, (3) bulkhead, (4) flexible wire for recirculation baffle, (5) drain hose, (6) drain hose for bulkhead connection, (7) power cord, (8) motor shaft. (Courtesy of Frigidaire.)

the unit has dropped, this would prevent the flow of the water in the proper direction. If water does drip from the back, the slinger ring might be out of order, or it might be a normal condition caused by high humidity.

noise

A certain amount of noise is normal in an air conditioner. The Freon will make a hissing sound as it travels through the system. If you pulled the plug you might hear it until the high side, or "head" pressure, had diminished through the restrictor. When operating, the "pulsing" of the compressor can set up a harmonic vibration in the tubing. If the vibrating tubing rubs on something, a "rattling" noise would ensue. (The tubing can probably be bent to stop the noise. If not, a piece of sponge rubber can be inserted to absorb the vibration.) A "thumping" noise heard when the motor runs might be traced to the motor mounts (see Fig. 2-18). Three things might be wrong: (a) some motor-compressors

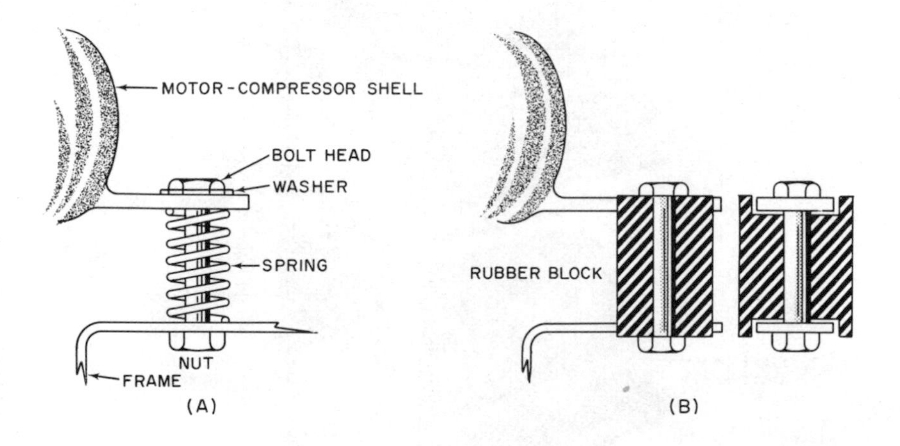

Fig. 2-18. Typical motor mounts.

are shipped with special bolts that should be removed and discarded—they might still be in place; (b) another type of mount has a spring and/or rubber arrangement with a bolt that should be loosened and left in place—the bolt might not be loosened enough—this is easily checked by pulling the plug and wiggling the motor; (c) in another compressor mount rubber blocks are inserted to absorb the vibration and hold the noise down to a reasonable level—here again, it might be necessary to turn the bolts that hold the mounts in place. Sometimes a little experimentation is required.

If the air stream increases greatly in velocity and its attendant noise becomes objectionable, the trouble might be excessively high voltage making the fans run faster. That is a rare occurence, but, if it ever happens, the power company should be notified at once.

Ticking noises are caused by the fan blades, but they can be bent so they will not hit anything. Scraping noises might be caused by the slinger ring. Here again, bending or adjusting will provide clearance.

A definite "squeal" might be caused by a dry bearing. If you do not oil it, the whole unit might become inoperative. When the bearing finally seizes, it will stall the motor. If the motor does not have an overload protector, the excessive draw of power will blow the fuse.

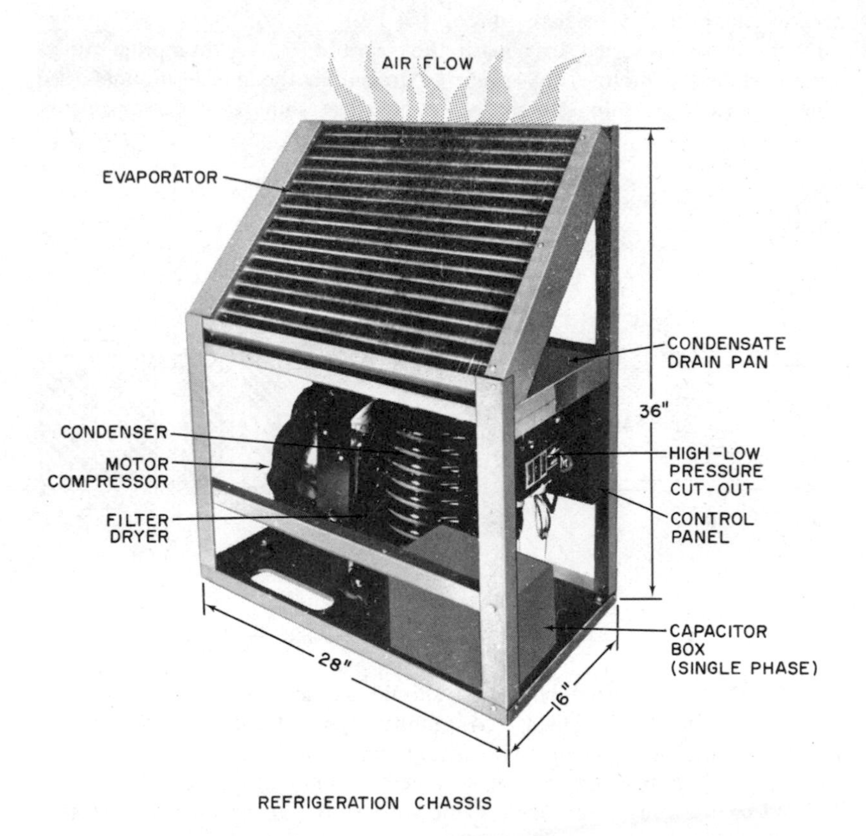

Fig. 2-19. Chassis of package unit. (Courtesy of Emerson Radio and Phonograph Corporation.)

Passing mention should be made of another type of condenser or evaporator. This type is often called a "coil," even though the term is a misnomer, because "coil" implies that the tubing is spiraled whereas actually, it just goes back and forth in loops. (You might find a spiral of tubing if it were in a tank of water). As far as window air conditioners are concerned, the term is a convenient one to use.

A "pin" coil is presently in use on which the fins have been replaced by a multitude of welded pins. The heat transfer of this type is increased 50%, but a cleaning problem would be encountered if the filter were left out of the unit. Dry dust would not be much of a bother, but certain vapors, especially kitchen grease, will quickly form a sticky film on the metal, and "seize" whatever dust comes along.

The "package" type of unit works on the same general principle as the window type, but it uses water to carry away the heat of the condenser instead of air. A package unit may have ducts attached, in which case it is part of a central system; if not, it is "free standing." A unit of this latter type is shown in Fig. 2-19. Water-cooled condensers present special problems, which we will consider later in the book.

The heavier equipment runs on 208/230 volts, but it is easier to learn the principles of operation by studying the 115-volt units first. The next three chapters will deal, mainly, with the difficulties of a 115-volt air conditioner with thermostatic control.

3—Unit Inoperative

The unit will not run, obviously, if it is not receiving electricity. The fuse might have blown or the circuit breaker might have kicked off. If the fuse has blown it does not necessarily mean that anything serious is wrong. Fuses sometimes get "old and tired" and blow for no other particular reason. With either type of safety device, if there are several appliances on the same circuit, and they all started up at the same time, something might "give." A motor needs more current to start than to run so that, if, by coincidence, they all started together, such an abnormal load could blow the fuse. Or, they might all run for years without that happening. If the power supply should fail for a time because of a storm or repair work being done on the lines, all the thermostatically operated devices might start at the same time, as well as anything else (such as an appliance) that happened to be in use when the power failed. In this situation, the excessive (and abnormal) power demand would most likely blow the fuses. When this happens shut off the appliances, replace the fuses, and start the appliances one by one.

If the unit will not run, it could be that the plug has fallen from the receptacle or that the prongs of the plug are not making contact. If it is in place pull the plug, spread the prongs apart (see Fig. 3-1) and reinsert it. Sometimes a plug fits too loosely, and, if the cord gets "jiggled," one or both of the prongs will loose contact with the outlet.

use a tester

If the unit still does not start, it will be necessary to use a two-wire tester (test bulb), which may be purchased at an electrical supply store, to locate the trouble. A tester can be improvised, however, which will serve well enough for the 115-volt circuits under consideration.

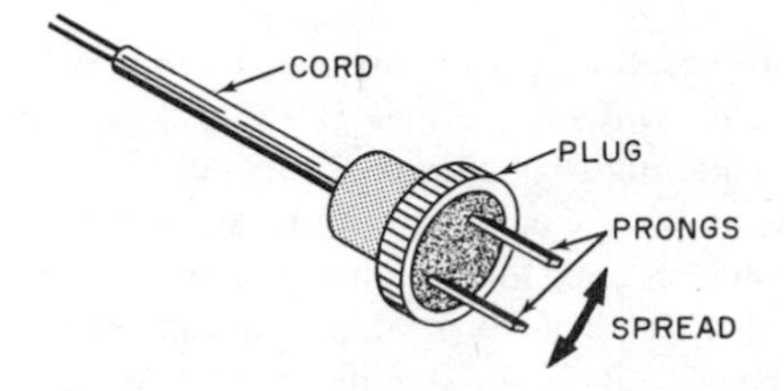

Fig. 3-1. Spread prongs of the plug for better contact.

To make such a tester, take an extension cord that has a socket and bulb, and remove the plug. Slit the combined wires and remove the insulation from the end. You will then have two exposed strands of wire to work with, although regular solid probes (see Fig. 3-2) are much better.

Insert the probes into the receptacle. If the bulb *does not* light, there must be a break in the circuit from there to the fuse box, or a screw has become loose and fallen out either in the receptacle or on the fuse-box mounting plate. (We are assuming that the fuses or circuit breakers are all right, of course).

If this bulb *does* light up, the next step is to determine if, with power at the receptacle, electricity is reaching the machine. One thing that might keep the power from reaching the machine is the use of extension cords to carry electricity to the appliances. This is not a particularly good idea, if for no other reason than that extension cords have caused many fires. Aside from that, every wire offers resistance to the passage of electricity, and the resistance of a thin wire, used in some extension

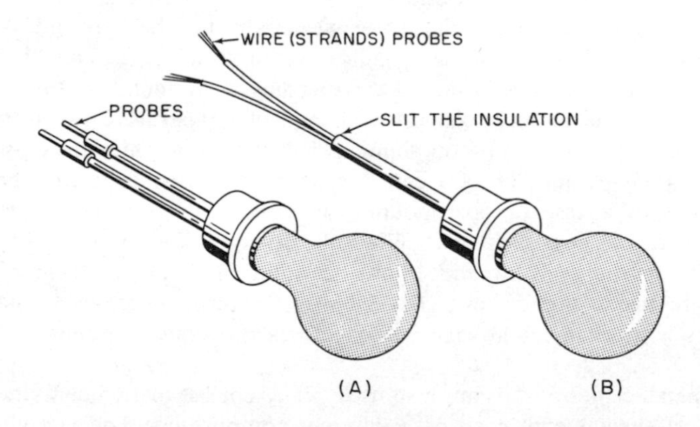

Fig. 3-2. (A) Test bulb with metal probes, (B) improvised tester.

cords, may be so great that the electric power will be reduced to a point where a motor will not run as it should. In this situation, it will not start readily and might overheat while running.

How do extension cords cause fires? Well, they are generally on the floor where they are stepped on, resulting in easily bruising or breaking the tiny strands of wire. The current arcing across broken strands gives off considerable heat, and, if the arcing is intense, the heat will burn off the insulation and proceed to char a wool carpet or anything else that is nearby and combustible.

To check the power input to the unit, it is necesary to remove the sheetmetal cabinet. This may be done by removing some screws. It is unlikely that you will have to use an Allen wrench or tiny screwdriver to get the knobs off; the cabinet generally has holes enabling the knobs to slide through. With the cover removed, and the controls and the control panel exposed, you will find the power cord leading to a terminal block. The cord should have a white wire on one of the terminal-block screws and a black wire on another one. If touching the test bulb probes to these two screws results in lighting the bulb, you can reassure yourself that the cord does not have a broken wire inside the insulation or a bad plug connection.

how a unit starts

At this point let us consider how the thermostat (see Fig. 3-3) starts the compressor. Remember that we are only thinking of a simple, 115-volt air conditioner with only one thermostat; the heavy, complex units have more than one temperature-sensing device. In fact, at this stage we can even forget about the fan.

On the terminal block, the white wire indicates the "ground" side of the circuit, and the black wire indicates the "hot" side. The "hot" terminal will have a wire going to the thermostat. Remember, the thermostat is a device that can convert a change of temperature to mechanical movement. It has a bulb containing gas that clamps onto the evaporator (see Fig. 3-4), and this gas can flow through a tiny tube to a bellows, which is in a case or box mounted inside of the unit. (It is probably secured to the control panel.) When the surrounding air becomes warm, the gas in the bulb expands, flows through the tube, and exerts more pressure inside the bellows. The bellows, in turn, increases in size and pushes a mechanical linkage, which closes the contact points (see Fig. 3-5).

Bimetallic thermostats are also used. They consist of two combined thin strips of sheet metal, each of a different composition. The warming will expand one more than the other, causing the two as a unit to warp (see

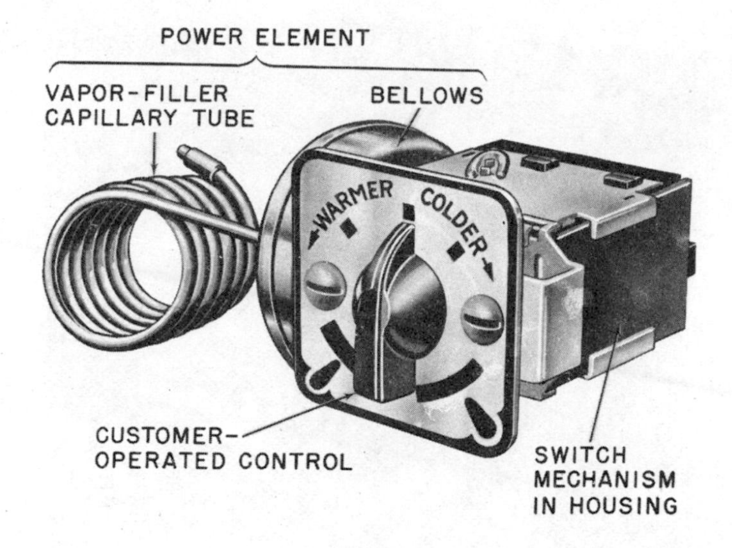

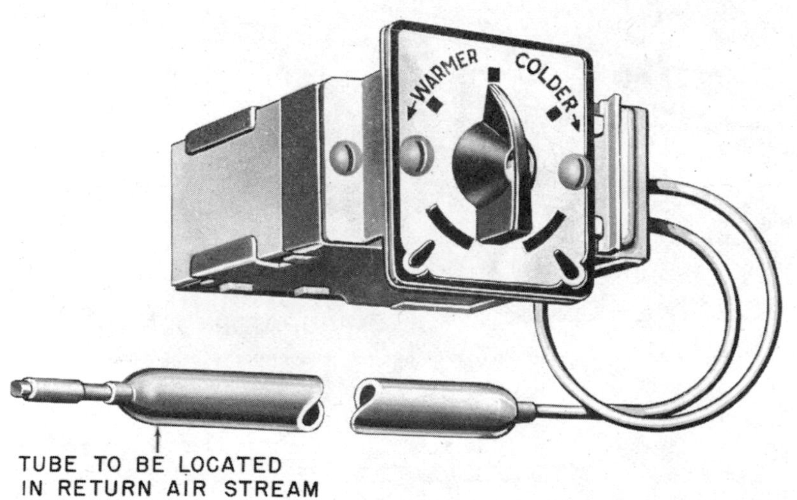

Fig. 3-3. Typical thermostats. (Courtesy of Ranco Inc.)

Fig. 3-6). Here again, the effect is to close the contact points. However, as far as the electrical circuitry is concerned, it makes no difference which type is used.

The thermostat is wired in series. That means that the current from the "hot" side of the input terminal block has to go through the contacts

Fig. 3-4. Thermostat bulb held against evaporator with bracket.

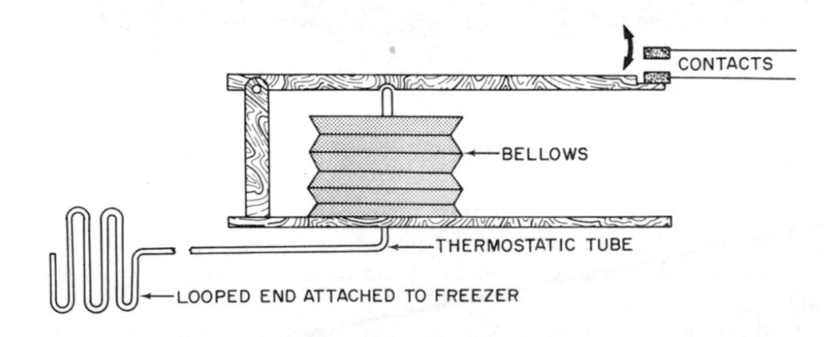

Fig. 3-5. Operation of a thermostat.

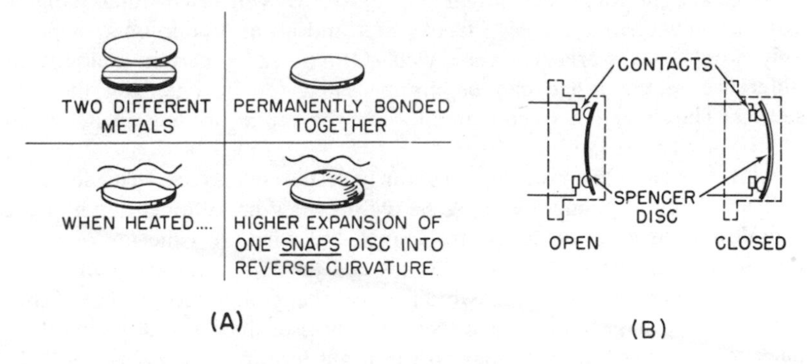
TWO DIFFERENT METALS
PERMANENTLY BONDED TOGETHER
WHEN HEATED....
HIGHER EXPANSION OF ONE SNAPS DISC INTO REVERSE CURVATURE
(A)
CONTACTS
SPENCER DISC
OPEN
CLOSED
(B)

Fig. 3-6. Bimetallic thermostat.

to complete its circuit, which is to the relay and on to the compressor motor. When the air in the room and the coil cool enough, the action is reversed—the gas in the bulb cools and the contacts open.

testing the switch

If you want to know if the power is getting through to the relay, with its direct connection to the motor, test the thermostat. To test any "make-and-break" device, simply bypass it with a wire. In this case, if everything else is all right, you still can start the compressor with the contacts open.

If the motor starts, there is no question but that the switch is defective. If you want to quickly restore the unit to operation, pull the plug, place a jumper wire between the terminals of the switch (see Fig. 3-7), and plug in the unit again. That will make the unit run, and you can pull the plug again if the room gets too cool.

We have assumed that the terminal connections are of the screw type; if they are of the "knife-and-blade" type, you can pull off the blades and clamp them together with a spring-type clothes pin. This procedure does not provide a very good connection, however, and the unit should not be left running this way for any length of time.

Replacing the switch is a straightforward mechanical operation. There is something you must be careful about, however. The little tube cannot stand abusive treatment. If the tube becomes flattened, closing the internal passage, the tube is useless.

If an exact replacement cannot be obtained, a different switch that can be physically mounted to the thermostat might work after a fashion, But, the various switches on the market have such different temperature ranges (on and off points) that, if a different switch were used, the thermostat's performance might be upset. If you tried interchanging switches between, say, a deep-freeze unit and an air conditioner, it probably would not work out very well. Of course, a certain amount ot difference in the range can be compensated for by changing the dial setting. The defective switch might have the range printed on it, which can be used as a guide.

At this point, the current goes through the switch or bypasses it if you are using a jumper wire. A heavy-duty air conditioner might have a multitude of wires on its control panel, but what runs the motor is the circuit through the switch to the relay and the two wires from the relay to the starting and running windings of the compressor motor. Also, there will be a ground wire on the compressor motor that might lead back to the white wire terminal on the input terminal block.

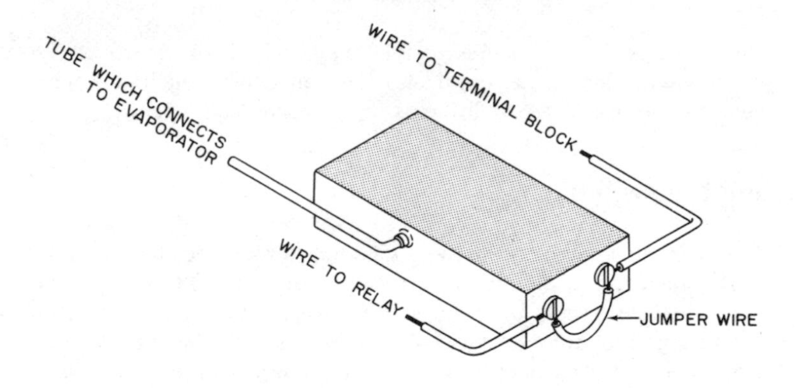

Fig. 3-7. Jumper wire on thermostat terminals.

Some people jump to the conclusion that, if the unit will not shut off, the relay contacts must have fused together. If you know the principle of operation, you will realize that, if the thermostat shuts off the supply of current, the compressor will stop, regardless of the condition of the relay.

Consider the situation if the thermostat bellows or bulb develops a leak. When the gas is partly gone, a greater change of temperature will be required to expand (or contract) the remaining gas in order to move the contacts. Thus, the unit will be running longer than normal, and shutting off longer than normal. Continued leakage will exhaust so much gas that the contacts will not move at all; they will just remain open or shut–more likely open.

the relay

The relay can energize the starting winding briefly because a motor draws a great deal more current when starting than it does after it reaches its rated speed. Any motor acts as a generator when it "gets going." This builds up a "counter flow" of current that opposes the flow of power from the fuse box. To be more exact, the rated-speed current is about 22% of the starting current. There are different types of relays, but in all of them the "high inrush" has the effect of moving a contact arm which closes the starting-winding circuit. When the "high inrush" drops off, the contacts open and disconnect the starting winding.

The circuit to the motor might contain one or more capacitors. They will be discussed later. There also will be a motor protector. Compressor motors always have a protector, although it will not necessarily be found on the fan motor. You can make a bypass test around a protector in the same manner as you did on the thermostat. These protectors are of the self-resetting type: if the motor heats up too much, the protector will open the circuit, and, when the motor cools off, it will close the circuit. The power flowing through the wire in the protector acts like an electric heater, and the heat causes mechanical action. Thus, in effect, it is a special purpose thermostat.

common electrical terms

In the next chapter, we will learn about reasons why a compressor motor will sometimes only "buzz" and not turn over, or at least not fast enough so that the starting winding will disconnect. One of the reasons is poor voltage. But, before treating that subject, it would be well for us to have a clearer definition of "voltage" and related terms. We have been using "voltage," "current," and "power" rather loosely, and, for purposes of diagnosis of electrical troubles, we should know the meaning of these words more precisely.

Power is measured in watts and, actually, is a measure of how much heat is dissipated by anything receiving electrical energy. Current, measured in amperes, refers to the number of electrons flowing past a given point in one second. It is apparent that, if the current in an appliance becomes greater, more heat will be given off. The relationship between current, power, and voltage, as well as resistance, can be stated in several ways by formulas.

Voltage is electrical pressure, or force. The current has to have this "force" to overcome the resistance of the wires. If the force is weak, or "low," the apparatus involved, such as a motor, will not function as it should. Low voltage is a widespread problem. It might derive from inadequate power-distribution facilities or from inadequate wiring inside the house. Also, power plants are limited in the energy that they can send out through the wires. This limitation becomes quite troublesome at times of peak demand. Power companies are using "standby" equipment, such as airplane engines, to provide the extra generating power during the hour or two of the way when the current demand is the highest. If they did not, there would be a considerable voltage difference over the area being served.

poor power supply

The problem derives from the fact that present-day usage of electrical equipment has quadrupled the requirements for power compared to the 1930's, when the requirements were not much different from what they had been for decades.

Consider this anology: If everyone on a street opened *all* their water faucets at the same time, the water pressure would fall and the main line under the street would not immediately be able to supply the increased water demand. In an electrical distribution circuit, when too much current is being "drawn off," the pressure or, as we say, the "voltage" drops. With a heating element, such as in a water heater, if the voltage is too low, it does not get as hot as it should. With a motor, on the other hand, low voltage might not permit it to start, and, if it does, it will run hotter then normal, which does not do it any good. In fact, you might say that a motor running on low voltage is destined to premature failure.

Air conditioners, in particular, are overloading the power circuits. A 230-volt air conditioner has two or three rather heavy motors. During a heat wave, they are running almost continuously. A washing machine, on the other hand, would be in use for a relatively short time.

The need for heavier lines increases continuously, but power companies are not to be able to string wires fast enough to compensate for the increased demand.

Aside from the limited funds power companies have available for modernization, many homes have the same small wiring that was put in them many years ago. And, it seems that they generally did not provide enough outlets, resulting in the use of extension cords which can cause a serious voltage drop. Even in newer homes, the occupants might have more appliances than the builder anticipated. No immediate, overall remedy is in sight, but manufacturers have made some motor improvements to help matters. The 1962 models are said to have a "very good power factor." That is another way of saying that the motor can make better use of the power supply that is available.

4—Hums, Will Not Start

Low voltage might be partly or entirely responsible for this condition. More exactly, either the compressor does not develop enough torque or something is preventing the compressor from turning as freely as it should. Or, it could be a combination of these conditions. Among the possible reasons for poor compressor torque are failure of the relay contacts to close, burning of the relay contacts, a defective starting or running capacitor, shorting of the running winding to the starting winding, and the shorting of either or both windings to the compressor-motor housing.

As for the matter of interference with proper compressor movement, the trouble might be sticky or stuck bearings, or excessively high pressure on the "high side."

testing the voltage

If a low-voltage supply is suspected, it is very useful to have a voltage tester so that you can determine exactly what the voltage is and how much it drops when the compressor motor is energized. If you do not have one, you still might be able to learn this information if the voltage is seriously low. In electrical work, there often are ways of arriving at conclusions by roundabout means. In this case, you can move a TV set to the house wiring circuit under consideration and observe whether the picture is smaller than normal. If it is, there is no question but that the voltage is very poor.

If the voltage input to the terminal block is adequate, there might be a voltage drop beyond it caused by loose connections or frayed wiring (see Fig. 4-1). Or the short might be inside the relay. The relay

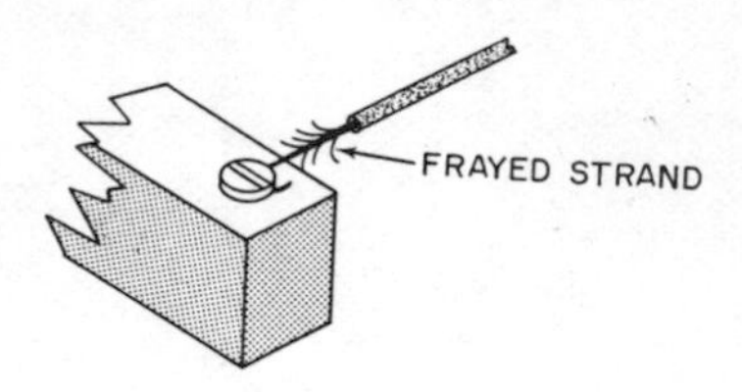

Fig. 4-1. Frayed copper strands will reduce power input and drop voltage.

has contact points similar to those in an automobile, and they also are subject to deterioration. When the contact point becomes pitted and burned, they cannot conduct the current necessary to start the motor (see Fig. 4-2).

relay failure

Why a particular relay fails is problematical. It might be that the bearings are sticky and the voltage is low with the result that, just after the points separate, the motor slows down, and the points close again. This situation might continue until the motor warms up (or cools off, as the case may be) and is able to maintain a constant speed of rotation. In this situation, the movable point might barely touch the other and move away. This erratic movement results in considerable arcing, and the heat given off damages the points. When the points become badly pitted and burned, the motor is not able to acquire enough speed to open the points; the motor will simply "hum" and, after a while, the motor protector will disconnect it from the power supply. This condition results from the windings drawing an excessive amount of power when the armature is in a stalled, or nearly stalled, condition.

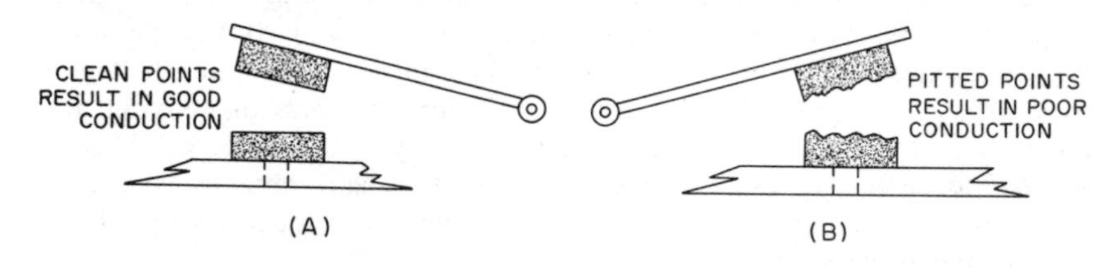

Fig. 4-2. Pitted and burned contact points result in poor conduction of electricity.

When you hear the "hum," try gently tapping the relay. This might make the points come together a little better so that the unit can start. But, this action remedies matters termporarily; the trouble will recur, and the indicated repair is replacement of the relay.

a faulty capacitor

Another common cause for starting failure is a faulty capacitor. The early light-duty models were made with a capacitor only on the starting winding. The heavier models were provided with another one on the running winding as well. In the process of improving the power factor, the designers were able to improve the motor torque to a point where the capacitors could be eliminated on some models. Most units still have them, however.

Starting capacitors are usually electrolytic and are wired in series to the starting winding of the compressor motor. Running capacitors are oil filled and are connected between the leads for the starting and running windings (see Fig. 4-3).

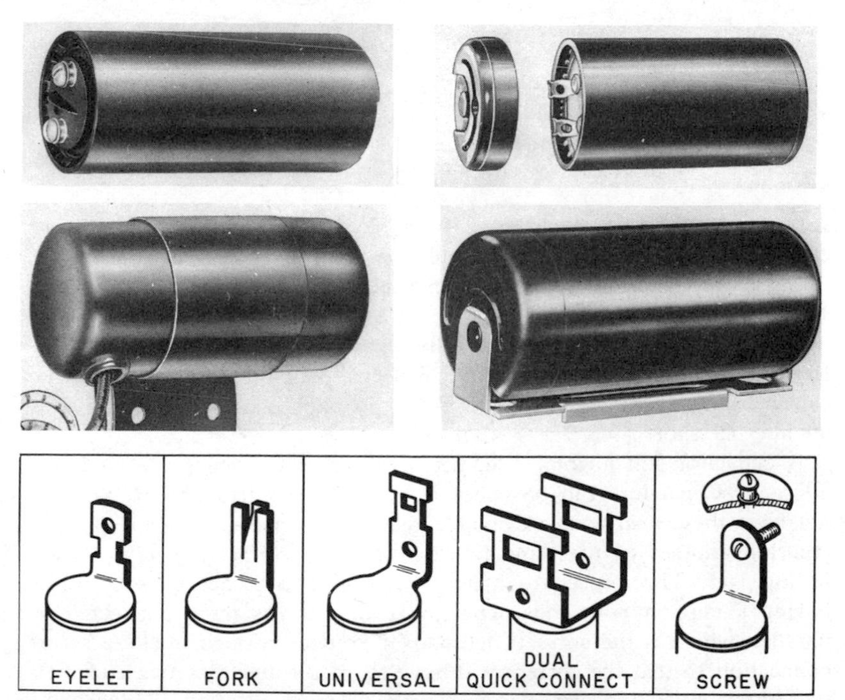

Fig. 4-3. Starting capacitors and typical methods of connection. (Courtesy of Cornell-Dubilier Electric Corporation.)

The capacitor is an interesting device; it has the ability to "store up" electrons and "give them back" the next time the current alternates. The effect of this is to improve the efficiency of the motor. The measure of a capacitor's storing ability is its "farad rating." A one-farad capacitor would be about the size of a ranch house. Practical capacitors are rated in millionths of a farad (microfarads).

Capacitors have some "weak points" that make them prime objects of suspicion when you have a will-not-start situation. The outside case, if it is plastic, can crack, and, if it is metal, can corrode. In either case, if the case opens, the electrolytic paste or oil can run out, thereby rendering it useless (see Fig. 4-4A). A capacitor cannot stand excessive

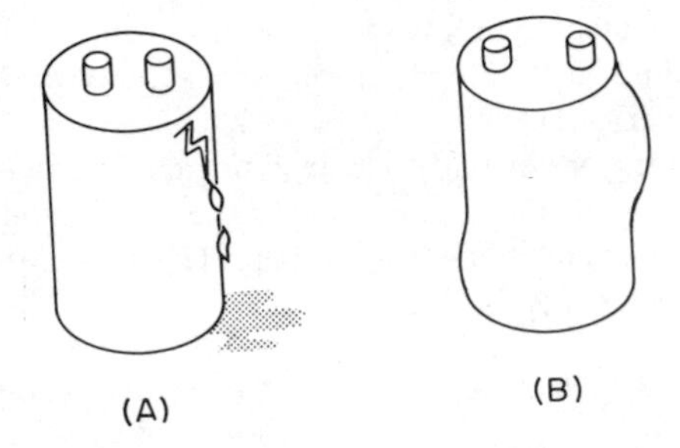

Fig. 4-4. (A) A cracked capacitor case will permit fluid to leak out; (B) a bulging capacitor case indicates an internal short.

voltage or too much heat, and, if an electrolytic capacitor was being energized every few seconds, it would be ruined before very long.

Capacitors deteriorate when not in use. If a unit has been shut off for several months and will not start up, it is very possible that the trouble will be traced to a capacitor. Carefully examine the case of the capacitor. If you find oil or paste oozing out, it is a positive indication of a defective capacitor. Or, if the case of an oil-filled capacitor is bulged, this indicates an internal short (see Fig. 4-4B).

A capacitor still might fail to perform properly even though it looks all right. Capacitors can be checked by certain instruments. If you do not have these instruments available, you still can check a capacitor by securing another of the same farad rating and connecting it as shown in Fig. 4-5. This procedure merely requires two pieces of wire, each having a clip on both ends. The illustration shows them connected in parallel, which is the correct method. Of course, you can make a series connection so that the current will have to go through first one and then the other, but, if the suspected capacitor is "open," it would "block off" the flow of current through it, and you would not learn a thing.

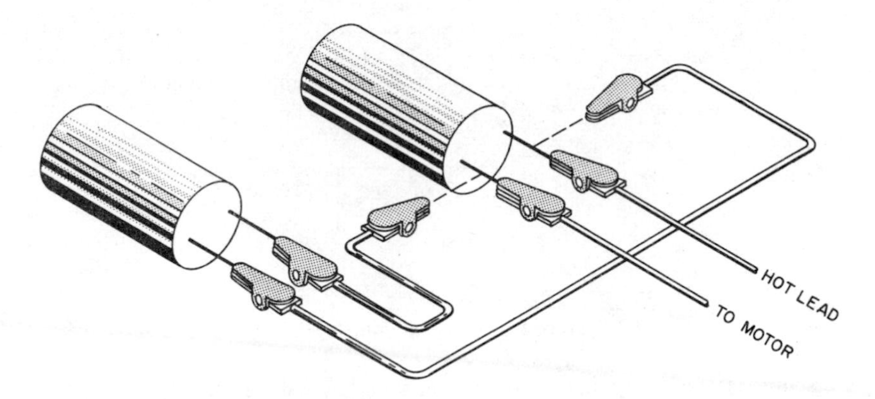

Fig. 4-5. Method of connecting capacitors in parallel.

To develop maximum torque, a motor has to have a capacitor of exactly the farad rating specified by the designers. A motor having a capacitor with a farad rating either more or less than specified will provide less torque, and it is quite possible that it will not start at all. For testing purposes, however, a capacitor having a farad rating somewhere in the general vicinity of the one being tested might work all right.

With some capacitors, there is the mattter of a-c polarity to bear in mind. With d-c capacitors, the terminals have to be connected in a certain way with respect to whether they are marked positive or negative (plus or minus). When the a-c capacitor was developed, it was designed so that it did not make any difference which way the two wires were connected. In recent years, however, a slightly different type has been specified for some installations. This type has to have the ground-side terminal connected as shown on the schematic for that particular unit. The terminal closest to the input from the switch is the "hot" terminal and the other one (identified by a spot of paint) is the ground, and, if you have the capacitor turned around so that the ground terminal is connected to the "hot" lead, the capacitor will not perform as it should.

a shorted motor

Another reason for starting failure is a short in the compressor motor. You might have a short circuit between the starting and running windings (see Fig. 4-6) or between a winding and the motor housing (see Fig. 4-7). In either case, the electrical balance is upset, with resultant poor torque and heavy draw of current. This condition causes the protector to

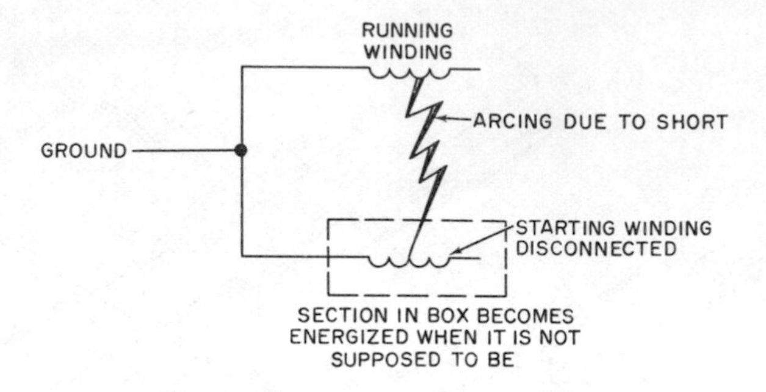

Fig. 4-6. Shorted windings.

disconnect the motor before it reaches rated speed, or shortly thereafter. Later, we will show how the motor can be tested to provide a fairly good indication of what actually happened. If the tests indicate that the motor itself is defective, little or nothing can be done about it. The entire unit, or at least the motor-compressor assembly, will have to be replaced.

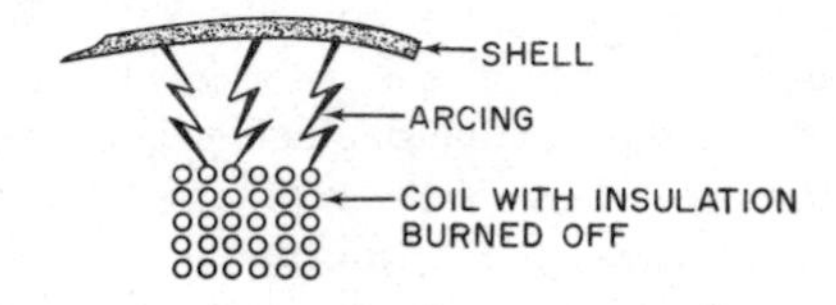

Fig. 4-7. A short to the motor compressor shell causes arcing.

Likewise, in the matter of sticky bearings, if the condition is too far advanced there is no remedy except replacement. Sticky bearings result from carbonization of the lubricating oil that is added to the Freon. (As a matter of fact, it also has an anti-carbonising agent, a water absorber and antifreeze.) Freon is not worth much as a lubricant so a petroleum product has to be added to it, and the presence of this petroleum is responsible for a large proportion of equipment failures.

lubricant breakdown

In refrigeration equipment such as air conditioners, the oil might serve as a lubricant indefinitely if it were not for the presence of moisture. Even microscopic particles of water cause a complex hydrocarbon reaction that results in the formation of a sticky compound that can adhere to the motor shaft. This "binding" of the shaft can progress to a point where the motor is unable to turn. The moisture, by itself, might not be too important if it were not for the heat which the water and oil are subjected to. A mechanism of this sort can get quite hot, and heat will accelerate these unwanted chemical reactions.

If the sealed unit is intact, we must assume that the moisture was in the air where the unit was manufactured. If the factories were in the Sahara Desert, it would simplify matters. However, they are located in relatively damp places, and the manufacturers take steps to dehumidify the air, and to bake the unit for a long time while holding it under a high vacuum. Even though advanced manufacturing techniques are practiced, there is always a possibility that a small amount of water vapor will remain in the system. If a unit has had six or eight years of use, it is likely that the bearings are at least a little sticky. It may be assumed, too, that the condition will get progressively worse.

high head pressure

If you encounter a hot motor that will not start, it does not necessarily mean that the motor has to be replaced. It might be that the heat has aggravated the sticking condition, and the shaft and compressor will turn when the motor cools. Actually, the condition is probably caused by high "head" pressure. The restrictor can pass only a limited amount of Freon, and, under certain circumstances, the high side, or "head" pressure can be quite high because liquid Freon is delivered to the restrictor faster than it can go through it on its way to the evaporator. If the unit stands idle for, say, 20 minutes, the refrigerant will have a chance to run through the restrictor and reduce the head pressure to relieve the stalled condition of the motor. Of course, the trouble might be caused by either hot bearings or high head pressure. In any event, the condition will usually remedy itself.

With these "will-not-start" situations, we have the classic pattern: the motor makes noises for a little while, perhaps five or ten seconds, until the "high inrush" or excessive draw of current causes the motor protector contacts to open. At that point, the motor stays "at rest" without electrical energy being delivered to it until such time as the protector points reset, which would be in a minute or two. Then the motor is energized again and the cycle is repeated. And that could go on until either the motor cooled off and started or you removed the plug.

bypassing the controls

It is apparent that the best way to determine whether the controls or the motor-compressor is responsible is to bypass all the controls and deliver power to the compressor motor directly. If the fan were wired to run with the compressor running, it would be taking power for its own use, which would, to some extent, reduce the voltage reaching the compressor motor. Aside from that, it would be time consuming to have to test all the wires and components on a complex unit individually to determine by elimination that the trouble is "apparently" in the compressor motor.

A three-wire test cord has been devised for this purpose (see Fig. 4-8). This tester has rubber-sheathed spring clips that can be attached to the three compressor-motor terminals. Such test cords are for sale in electrical supply stores. A professional test cord has a spring-loaded pushbutton that can be used to energize temporarily the wire leading to the starting-winding connection. If you wish, you can make your own (use heavy wire) with a toggle switch if you do not have a pushbutton switch handy. And, if you do make one, you will get along better with it if you fix the clips so that they can be attached at 70 or 80 degrees *away* from the perpendicular to the motor housing. In the process of making units more compact, compressor motors have been installed in ways that do not make the terminals particularly easy to get at.

A factory-made cord should have the function of the wires indicated by color. By convention (national agreement) a white wire indicates a ground connection, a black wire indicates the "hot" lead, and red indicates the wire for the starting winding. This should hold true as far as the test cord is concerned; if the mechanism being tested has a few dozen wires, they might have many different colors.

using the test cord

Now, what you must do with the test cord is to duplicate the action of the relay while providing an electrical supply direct from the receptacle. Here is how this is done: first, you must learn which terminal is which. You can determine this from the schematic, which should be pasted inside the cabinet somewhere. You *must* have the connections right. Do not guess. If the schematic is missing, defaced, or unclear, call the dealer. When you know which terminal is which, remove the wires from the terminals of the compressor, or, if it has a capacitor, from the capacitor input terminals. If you think you might not remember where they came from, make a drawing, and use bits of colored tape or tags to identify them. Attach the black-wire clip to the terminal for the running winding (*do not* have the test cord plugged in), and attach the red and white clips to the starting and ground terminals, respectively. When the test cord is plugged in, push the button to allow some of the black-wire current to flow in the red wire and energize the starting winding (see Fig. 4-9). Push the button for a few seconds (five seconds is the maximum). If the motor does not start in five seconds it is not going to, or at least not until it cools.

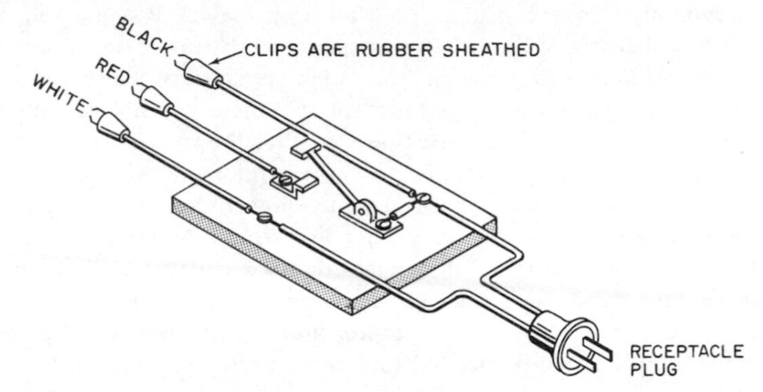

Fig. 4-8. Three-wire test cord.

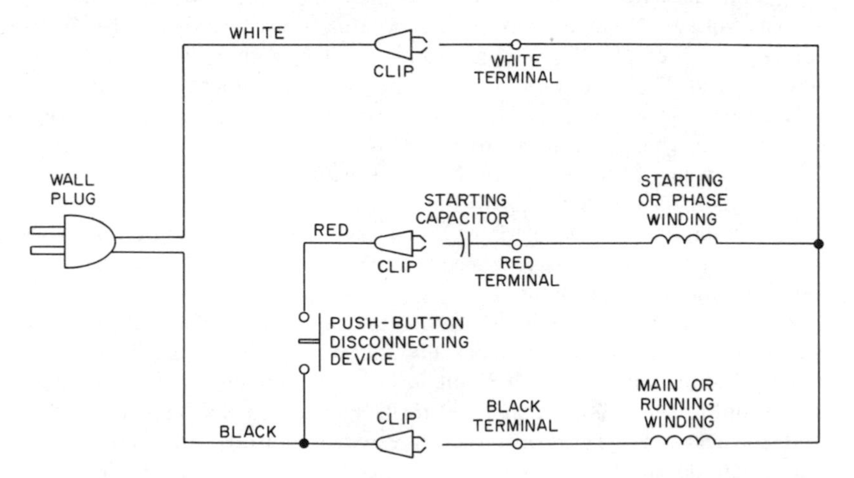

Fig. 4-9. Motor with three-wire test cord connected.

In properly made cord, the black-wire current will run through a fuse between the pushbutton and the plug. A stalled motor will blow a 15-ampere fuse. If the motor starts, and you interchange the black and red wires when you put them back, you then will have the relay disconnecting the running winding, leaving the starting winding to keep the motor going. That is not the way the motor was intended to run, and it will operate at higher than normal temperature. If you put the red wire on the ground terminal, on the other hand, when the relay disconnects, the motor will have an incomplete circuit and will stop.

In the interests of economy of operation, cooling equipment has compressor motors that are perhaps a little small with respect to the torque required to start the compressor. A larger motor could be used, but the monthly power consumption would be higher. Because the consumer insists that the equipment be as economical to operate as possible, we have the situation in which a unit might operate properly for some years; but, when the bearings become "sticky", there will not be sufficient motor torque to overcome the stickiness so that the motor will start and run reliably. It even could happen that the equipment will not start at the times of the day when the community power demand is high.

Can anything be done? Yes, very possibly. Where the voltage delivered to the unit is poor because of thin wires or poor connections, correcting these difficulties might keep the unit in operation for several more years. Heavier wires, preferably on a separate circuit from the fuse box with the terminal ends properly cleaned and the screws well tightened, might suffice.

If that does not do it, a special transformer might be used to step up the voltage. Whether or not you do this, however, should be determined by a person who is completely familiar with power distribution and motors. The "boost-and-buck" type of transformer is intended for this service. Such a transformer is designed so that the various windings can be interconnected to provide a great variety of output voltages. They can stand 24-hour use without overheating.

The "tapped-winding" transformer does not cost as much, but it has little flexibility. If the relationship of the input and output voltages happens to coincide with your needs, well and good, and if it does not, nothing can be done about it.

Sometimes the "low-voltage" difficulty stems from the fact that (to speak of the heavier equipment) two different voltages are being supplied. In some situations a 208-volt input is being supplied to equipment that should have 230 volts supplied to it. The reason that tapped-winding transformers were placed on the market was to convert 208 volts to 230. They can do so, but, if the input is not even 208, or if it fluctuates considerably, the "boost-and-buck" type is more useful.

Incidently, it is remotely possible that the reason for the low voltage is that the power company has its distribution taps inproperly connected. In that (unlikely) event, they would remedy the situation.

Sometimes fans will not start. They, too, have capacitors. Here again, an exact replacement is advisable. In the oil-filled type, there is a voltage rating as well as a farad rating. If you get a replacement made by the same company, it will have the same physical dimensions and you will not have to alter the mounting bracket to use it (see Fig. 4-3).

5—Runs, Will Not Cool

low voltage

Here again, low voltage might enter the picture. If the unit is started when the power demand of the area is fairly low, it will operate at proper voltage. As hours pass, the power demand might rise, causing the unit to run on a lower voltage. For example, if it is started at 9 A.M., that would be a time of the day associated with low household-power consumption. If it is still running at 11:30 A.M., a voltage drop would be encountered because people would be preparing meals and their ranges, electric fryers, and such would be in use. The low voltage condition, if serious, would slow down the motors.

As was mentioned, the unit cannot cool if the air is not moving normally. This statement applies to the movement of air through the evaporator coils and to the flow of air carrying-away the heat of the compressor and condenser. If either fan stops, the room will warm up.

What we will concern ourselves with directly at this point, however, are reasons for the temperature of the evaporator coils not being as low as it should. This situation might stem from a restriction (obstruction) in the system, too much or too little refrigerant, air in the system, or a dirty condenser. Needless to say, having the hot sun shining directly on the condenser does not help matters either; the ability of a condenser to give off heat is dependent to the difference in temperature between it and the hot gas inside, and the heat of the sun reduces the differential.

In addition to the foregoing, you might have a condition in which a malfunction of the compressor (stuck or broken part) results in the pressure equalizing on both sides of the system. In other words the pump was not pumping.

restrictions

Suppose it was a restriction; this situation is similar to the condition described in Chapter 4 in which "sticky" bearings resulted from carbonization of the oil. Actually, when the oil carbonizes, minute particles are circulated with the Freon, until they finally adhere to some part. In this case, they adhere to the walls of the restrictor.

What is this restrictor? Recall that we said earlier that the liquid Freon has to be held back, or restricted, so that the drops have a chance to vaporize? In small units, this function is accomplished by having the refrigerant flow through a length of spiral tubing so small that its internal diameter is measured in thousandths of an inch (about 0.035 inch). When the refrigerant flows through this tube, its flow is "restricted" by the friction developed between the fluid and the walls of the tube. Thus, the size and length of the tube used permits the right amount of Freon to flow through it, unless something reduces the size of the internal passage. When an impurity gets caught in the restrictor, microscopic particles of carbonized oil build up on the walls and around the impurity until the Freon flow is stopped (see Fig. 5-1). You might have

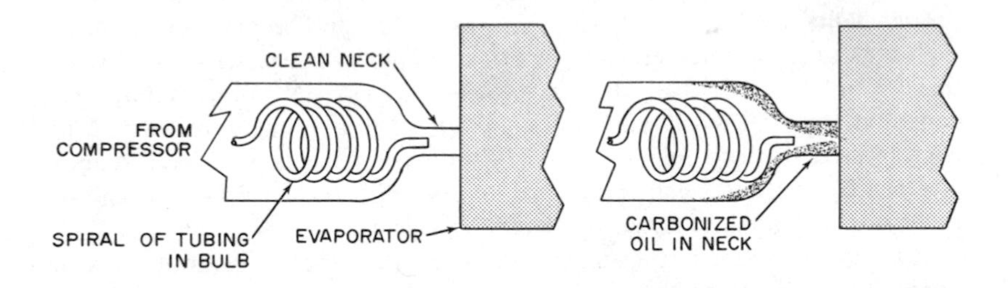

Fig. 5-1 Clean and carbonized restrictors.

a restriction at other points in the system, but the restrictor is the most likely place for it. A serious restriction usually calls for the replacement of the unit.

On the other hand, if the passage is clear but the system has too much refrigerant flowing through, the performance will be poor also. The evaporator will fill up, or "flood," and the Freon will be vaporizing in the so-called "suction" tube, which is the larger of the two tubes between the evaporator and the compressor. In this situation, you can detect the condition by sight and touch. If the weather conditions are right, the suction tube will actually frost up, or it will at least be cold and collect moisture.

too much refrigerant

How could the system have too much refrigerant? It could have been accidently overfilled when it was being repaired, and it is not inconceivable that it was overfilled at the factory. On some models, the refrigerant can be drained from the system by loosening a hex nut on the compressor "charging stud" (see Fig. 5-2). This condition can exist for quite sometime before it requires investigation, that is, at times when the unit runs for only a few minutes and shuts off, only a moderate

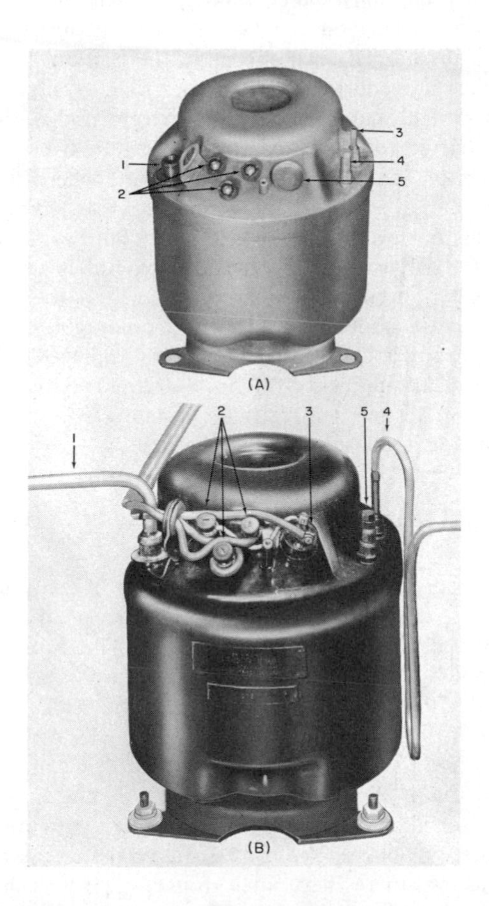

Fig. 5-2 (A) Motor compressor before installation: (1) suction line fitting, (2) protector holder. (B) Motor compressor after installation: (1) suction, liquid lines, (2) terminal cap nuts, (3) motor protector, (4) line to condenser, (5) charging stud. (Courtesy of Frigidaire.)

amount of refrigerant is pumped into the evaporator, and it does not flood. By he same token, it follows that an air conditioner could serve well enough during several weeks of mild weather and then give trouble during a hot spell.

Putting in refrigerant is called "charging" the unit, and, when it leaks out or is drained, the unit is said to be "discharged." On a partly discharged unit, you can see an incomplete frost pattern on the evaporator, or you can feel the coils on the turns not covered by fins, to determine which are cold. As the leakage progresses, two-thirds of the evaporator will be cold, then one-third, and, finally, it will not be cold at all. If all the turns are warm, you can not tell without instruments whether the refrigerant is discharged or if there is a restriction.

There is another procedure that might be tried. Applying a little heat to the restrictor might enlarge the tubing to dislodge the restriction. The spiraled restrictor tubing is contained in a bulb in the small tube from the compressor (see Fig. 5-1) at the point where the compressor is connected to the evaporator. You must be exceedingly careful in applying the heat, however, because the restrictor has soldered connecnections that can easily melt. This method sometimes works, but if the "slug" of dirty oil is permitted to circulate around the system, it will, sooner or later, cause another restriction. You might ask, "Why can't the dirty oil be drained and fresh refrigerant replaced?" This is done on heavy commercial equipment, but, on household units, the compressor clearances are so small and the restrictor opening so much smaller still, that no practical method has yet been devised.

the halide leak tester

If the unit has discharged, it is obvious that there must be a leak somewhere. Any of the soldered joints can leak, or the charging stud (sometimes called a valve), or the terminals. The tubing itself might develop a "pin-hole" leak, usually on a sharp bend, where the deformation resulting from the bending has weakened the wall.

When refrigerant leaks, the lubricant usually leaks with it. If you can find the spot where oil seems to be running out, this can lead directly to the leak; at least this is the case in a refrigerator. On an air conditioner having a fan motor that requires periodic oiling, you might not know for certain whether or not a spot of oil resulted from a leak or from oiling the fan. With the fan, there is so much moisture (from rain water), and turbulence of the air, that oil is easily distributed.

Nevertheless, if a leak appears, what is the testing procedure that can be used to discover its source? Probably the most common testing tool in use is the halide, alcohol-fired, flame-type leak tester (see Fig. 5-3).

Fig. 5-3. Testing for leaks with a halide tester.

This device draws air, needed for combustion, through a rubber tube. Placing the end of the tube at a leakage point, you will observe that a small leak changes the flame color to lavender, and a large leak changes it to green.

To put the tester into operation, you must fill the pan (located above the plastic base and below the combustion chamber) with alcohol and set fire to it. After a while, the combustion chamber will get hot enough to maintain a flame, using as fuel the alcohol fumes that rise from the container located in the base.

The flame will react with Freon in the manner indicated above. It also reacts with other materials, and therein lies some of the difficulty in leak testing. In the first place, if the alcohol fumes get into the air, they might discolor the flame and give a false reading. In addition, domestic fuel gas and some other fumes can cause little variations in the flame color. Some servicemen think they can distinguish one air impurity from another by the effect on the flame, but the variations are small and difficult to detect, even by an "expert."

What makes leak testing so difficult is that the Freon fumes (or if you spill the alcohol, the alcohol fumes) distribute all over the room. To test an air conditioner, you will have to shut it off or disconnect the fan, because the movement of air, besides making the flame flutter, will carry away the Freon fumes from the point of leakage. Under certain circumstances, you can have a situation in which the flame reaction would be akin to holding a geiger counter and approaching a radioactive particle—the nearer you get, the stronger the reading. That is of no help in localizing the leak precisely.

Thus, when you "fire up" the tester, you might contaminate the air all over the room, and you would have to air out (more-or-less replace the air) in the room, which is exasperating, and time consuming. And, while you were doing it, the flame would be using up the alcohol in the base container, with the result that the fuel could go out before you could finish testing. What is to be done? One procedure that will help greatly is to secure a Bernz-a-matic (or similar type) torch that connects to a two-pound tank of propane gas. The Bernz-a-matic torch can be used to heat the testing torch so that a flame will remain without burning alcohol in the priming pan. If you heat the test torch at a point some distance from the unit being tested, and have a ventilated room, you should be able to proceed with the testing. Another, more recent, flame tester is fueled by bottled gas. It is claimed to be "considerably more sensitive than the alcohol type." There is an electronic tester available, but it is so expensive that only the larger repair shops have them.

A tubing leak can be soldered, if it is located where you can get at it. Compressor terminal leaks generally can be corrected by tightening the terminal nuts. A charging-stud leak can be remedied by tightening the bolt on the end of the stud.

Where applicable, an excellent leak testing device is a three-dollar, war-surplus stethascope, with which to listen for the characteristic "hissing" sound. You can not have any fan noise, of course; the pitch is nearly the same as that of the hissing. The "pulsing" of the Freon makes a low-pitched rumble, and, where there is a leak, the different pitch of this sound can be detected easily. This idea can be used where there is considerable leakage, but, if the unit is badly discharged, you might have to add more refrigerant to produce enough internal pressure for the leakage to be detectable. Recharging methods will be described in a later chapter.

A shortage of refrigerant will change the normal operating pressures. The evaporator will not be able to deliver vapor fast enough to satisfy the capacity of the pump. In other words, the compressor will not "know" that the refrigerant charge has been reduced, and it will keep running at normal speed with the result that the "low-side" pressure will drop from, say 50 psi, to 30, to 10, and, after a while, it will try so hard to pump enough vapor to maintain the normal amount of liquid on the high side that it will actually develop a low-side "vacuum." If you had a leak on the low side, after a while, air will be sucked into the system. This is a very bad situation because the added moisture, and the air will carbonize the oil. In addition, the compressor will work very hard to pump the air. Thus, air in the system is characterized by an exceedingly hot condenser.

malfunctioning compressor valve

Another reason for little or no cooling is the malfunction of whatever serves as a valve in the compressor. The manufacturers have various notions about how compressors should be built, but they all have to have an arrangement to prevent the vapor from flowing back to the low side while it is being compressed.

One model uses a solid, off-center wheel inside of a cylinder (see Fig. 5-4). The wheel has a "divider block" bearing against it. This block slides in a groove and has a spring that holds it against the wheel. The eccentric mounting causes the wheel to move toward and away from the cylinder wall at the point where the divider block is located. The movement of the wheel alternately covers and uncovers two openings or "ports." One of these ports leads to the low side, the other to the high side. When the low-side port is uncovered, the heat-laden vapor from the evaporator flows into the space between the eccentric wheel and the cylinder wall. Then, the rotation of the eccentric wheel pushes the vapor against the divider block and squeezes it out through the high-side discharge port, thereby condensing it.

If the divider block cannot move freely in the groove because oil has adhered to it, it will be pushed to the end of its travel and stay there. You then would have a situation in which the piston moves air in and out through the valve openings, but no compression is taking place. This results in equalization of pressure on both sides of the compressor. A situation with similar symptoms can occur on a heat-pump model when the reversing valve leaks.

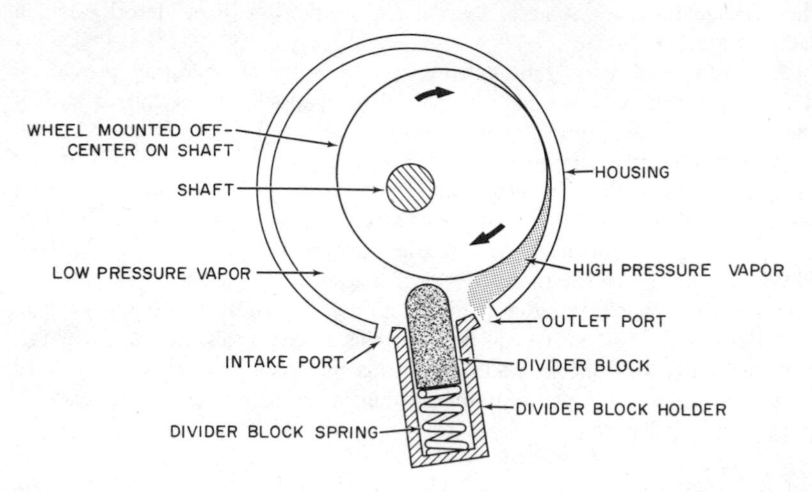

Fig. 5-4. Divider-bock compressor: (1) shaft, (2) wheel mounted off-center on shaft, (3) housing, (4) divider block, (5) divider block holder, (6) divider block spring.

a dirty condenser

If the condenser becomes covered with dirt (see Fig. 5-5), that, too, can nullify the cooling action. The compressor delivers its compressed vapor to a condenser, and, if the condenser stops working, the pressure will build up even though the refrigerant will bleed through the restrictor.

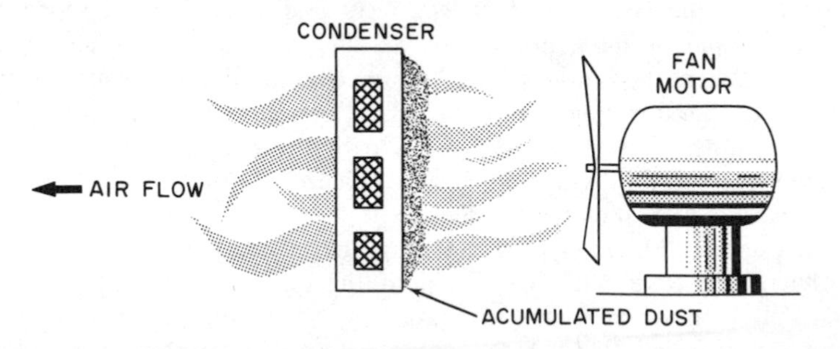

Fig. 5-5. Dirt on the condenser will prevent it from cooling Freon refrigerant.

When a coil becomes covered with dirt, it is often said to be "loaded up," and this confuses people who do electrical work because they automatically think of the term in an electrical sense. Actually, in a refrigeration unit, you can "load up" a coil in three different ways: a coil of wire, such as a solenoid, can be loaded electrically; a condenser or evaporator coil can be, on the outside, physically loaded with dirt; and a condenser or evaporator coil can be loaded up on the inside with liquid Freon (implying that it was "flooded," with little vaporization taking place).

6—The Conventional Refrigerator and Freezer

The common household refrigerator that has been in use for the past 25 years does not require a description of its general appearance or capabilities. In recent years models have been marketed that have little or no frost.

These models have a different principle of operation, which will be described in the next chapter.

air circulation

The early refrigerators with a "freeze box" at the top, cooled the food entirely by convection. That is, the air that touched the freezer would chill, become more dense, and settle to the bottom. This dense cold air would displace the air at the bottom, pushing it along the sides forming a continuous circulation pattern (see Fig. 6-1).

This considerable movement of air, combined with the frost gathering tendency of the freezer, tended to dry out the food. To combat this tendency, glass-covered hydrator trays were put in the bottom to conserve food moisture in such things as green leafy vegetables.

Then it was decided to put the evaporator coils in the walls so that the walls of the food compartment could cool by radiation. Many ideas were worked out with respect to coil placement and air circulation; the outcome was that the "across-the-top" freezer was introduced.

Here, again, we have convection, but, in this case, the chilled air flows down outside the food space. Actually, it drops mostly along the inside of the door (see Fig. 6-2). Although warmer, moisture-laden air will have to make its way up to the top and around the freezer, much

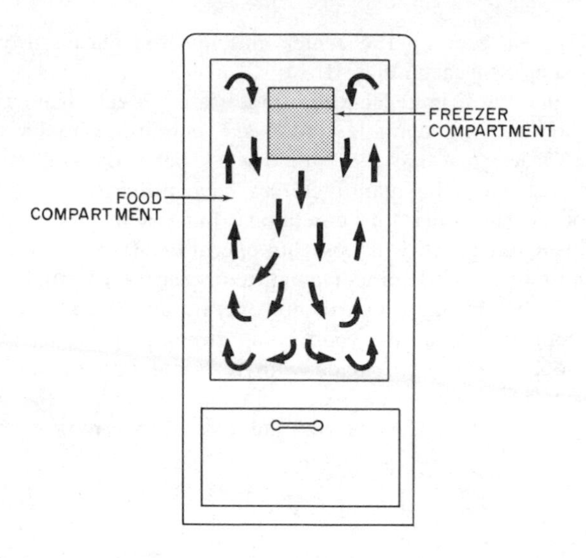

Fig. 6-1. Air circulation pattern in old-style refrigerator with center freeze box.

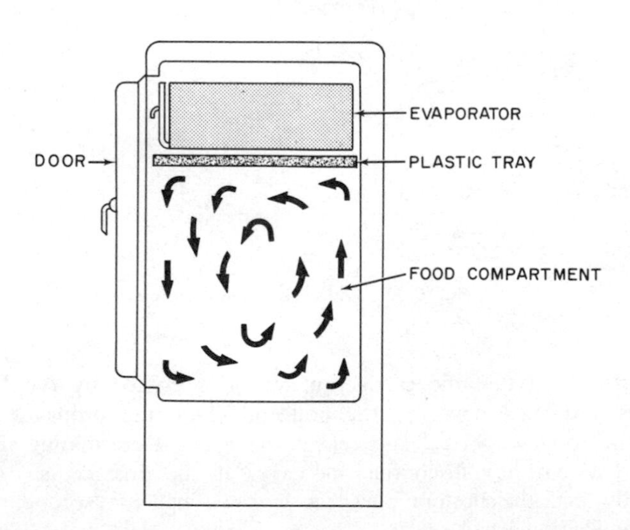

Fig. 6-2. Air-circulation pattern in refrigerator with "across-the-top" freezer.

of the warm air goes up the center and hits the plastic tray, which is not cold enough to accumulate frost.

The designers anticipated certain undesirable effects from the convection, but the varying climatic and usage conditions make it virtually impossible to design a unit that would work perfectly in every situation. In certain instances, the manufacturers have provided a baffle to compensate for unusual operating conditions, but placing the unit in a cold room or on a sun porch will upset its operation. If the "cold control" is set to maintain a food-compartment temperature of 40° F., and the outside air is 40° F., the unit will not operate and the freezer will attain the same temperature as the food compartment.

Some models have been made with a "secondary system" (see Fig. 6-3)—a separate sealed system of tubing in two parts—the top part of which is connected to the freezer and the bottom part is in the food

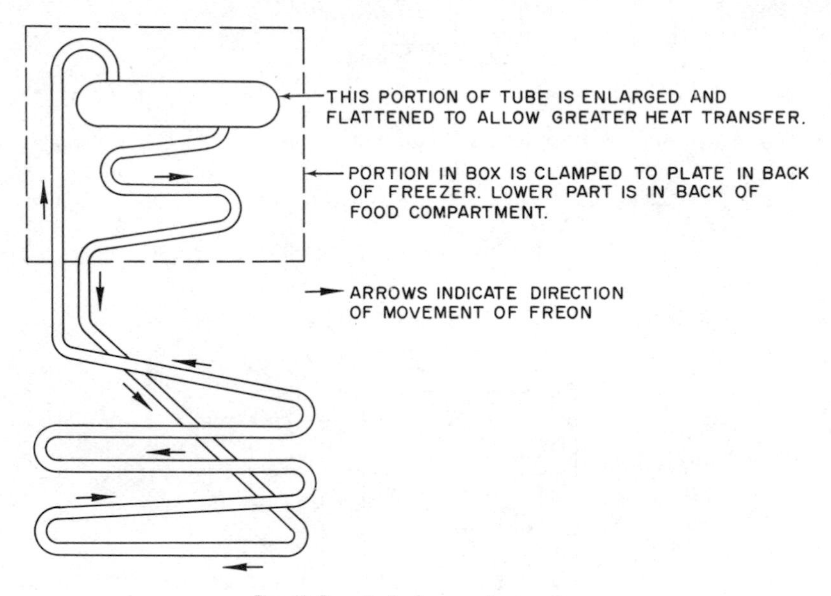
THIS PORTION OF TUBE IS ENLARGED AND
FLATTENED TO ALLOW GREATER HEAT TRANSFER.
PORTION IN BOX IS CLAMPED TO PLATE IN BACK
OF FREEZER. LOWER PART IS IN BACK OF
FOOD COMPARTMENT.
ARROWS INDICATE DIRECTION
OF MOVEMENT OF FREON

Fig. 6-3. Typical secondary system.

compartment. When the refrigerant vapor is cooled by the freezer, it liquifies and runs down to the bottom. There the drops absorb heat from the food, vaporize, and repeat the cycle. Determining if the secondary system has discharged is easy. If the freezer is functioning normally and the bottom part has warmed up, the secondary system must have discharged.

loose door seal

With either type of freezer, a leaking door seal can cause longer-than-normal running time and excessive frosting. A leak can be due to rotting of the seal (probably caused by fruit juices), flattening due to normal deterioration, a sprung door, or a loose latch. You can check tightness of the seal by taking a dollar bill and closing the door against it; you should feel some tension when pulling the bill out.

If there is generalized looseness caused by flattening or wearing of the door latch and striker (the stationary part), it can be corrected by loosening a few screws, moving the striker and retightening the screws. If the door is sprung, you might be able to put it back in shape by pulling on the top while bracing the bottom with your knee. Some of the bigger doors have tie rods inside them that can be adjusted. It is quite a lot of trouble to get at them, however; you would have to remove the door-liner panel.

If the seal has to be replaced, it can be removed by taking out some screws. Probably several hundred varieties of seals have been marketed (see Fig. 6-4). There are several methods of attachment, but, usually,

Fig. 6-4. Some of the typical door seals used on refrigerators.

the screws are underneath the semicircular part. The larger cities have supply houses that buy the more common seals in thousand-foot lengths. You can order the length you need, but it takes plain-and-fancy figuring to cut the corners for proper installation.

Because seals are made of flexible material, they do not provide even pressure all around the door. This does not matter with the older models because you can adjust the striker so that the door closes tightly. With the newer models that have magnetic door closure, a "thin" spot will cause a leakage of air that, around the freezer at least, will produce an accumulation of frost. The remedy for that is to insert shims behind the seal, building it up so that it provides even pressure. It does not matter what is used—even cardboard shims can be used, provided it is thin cardboard. Measure the distance between screw heads, then cut the shim(s) so that they can be placed between the screws or on both sides of a screw.

erratic lights

Cabinet lights sometimes act erratically, even when a new switch is installed. This condition is frequently found when a common household bulb is substituted for the original one. Some people gather that the trouble is in the socket, and, since it looks as if replacing the socket would require pulling out the food compartment (technically, the cabinet liner) to install a new socket from the rear, they just put up with it. Strangely the light might be out even with everything in perfect condition, including the bulb.

Refrigerator light sockets have a rubber sheathing that extends above the metal part of the socket (see Fig. 6-5). The purpose of this sheathing is to prevent moisture from getting through to corrode the metal. The specially made bulb that was furnished with the unit had a stem long enough to make contact with the terminal at the bottom. It has four full threads, while the ordinary bulb has about 3½ threads, and the

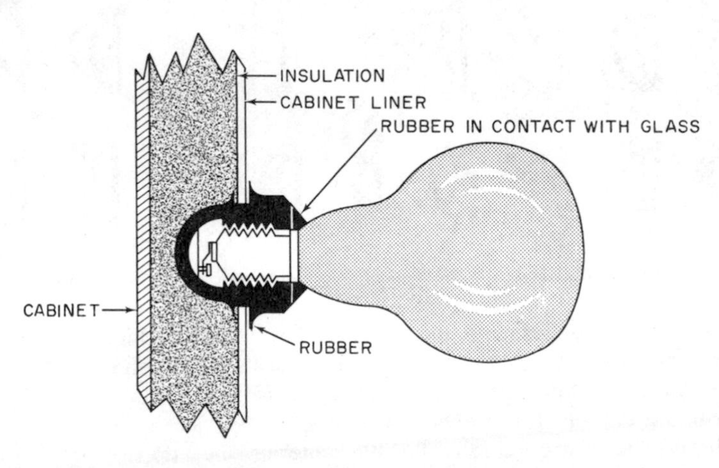

Fig. 6-5. Refrigerator light socket.

glass is likely to bulge out where it meets the stem. When this type of bulb is inserted, the bulge of glass hits the rubber, preventing the end of the stem from making proper contact with the terminal. And that is why the light flickers.

a faulty latch

How would you open the door if the handle came off? At the worst you could remove the hinges, which should be easy. Some hinge butts may be sprayed with a coat of paint, but usually they have a sheet-metal cover over the screw heads. These covers fit the butt castings perfectly, and stay on because, on the top and bottom, there are little indentations that fit into depressions in the casting beneath. There should be a turned-up edge so that a knife or small screwdriver can be inserted for removal; they snap back into place.

The door closure is in two parts. One part, the latch, "latches" onto the striker which is secured to the cabinet frame. Actually, the terms "latch" and "striker" are used interchangably, and, therefore, you should exercise care when ordering one of these parts so that you are not misunderstood and receive the wrong part.

When you pull the handle, the base of the handle presses against a plunger that, through a linkage, moves the latch up or down or sideways to disengage the striker (see Fig. 6-6). Normal wear will make it necessary to pull the handle back more and more before the slack is taken up and the desired action of the latch takes place. Washers or spacers often can be installed to compensate for the wear. If the handle breaks, the easiest way to open the door is to remove the handle mounting bracket and manually push the plunger. To get at the bracket pry the sheet-metal covering of the handle, thereby exposing the screws of the bracket.

If the linkage is broken, this idea will not work, because the hole in the door is too small to reach through with tools.

With an "up-and-down" closure, an enlargement on the end of the latch runs between two nylon rollers (see Fig. 6-7). One roller is fixed and the other moves against a spring. To open the door, merely slide a table knife between the cabinet and the rubber seal, and exert pressure as required (usually downward), while slightly pulling the door.

It is possible for the striker assembly to jam. Before sawing off the latch with a bayonet-type hacksaw, try removing the hinges and reaching in from the hinge side to smash the plastic molding covering the screws holding the striker. Replacing the plastic molding is simple compared with replacing the latch. A door liner has numerous fasteners and the linkage has tape and insulation around it. When the liner is removed, the insulation has a tendency to move, and you will have to keep the insulation in place while you adjust the liner for fastening.

If a "sideways" closure linkage breaks, you need only remove the hinge screws on the cabinet part; the other stays fastened to the door. Once this has been done, it is a simple matter to move the door sideways to disengage the latch from the striker.

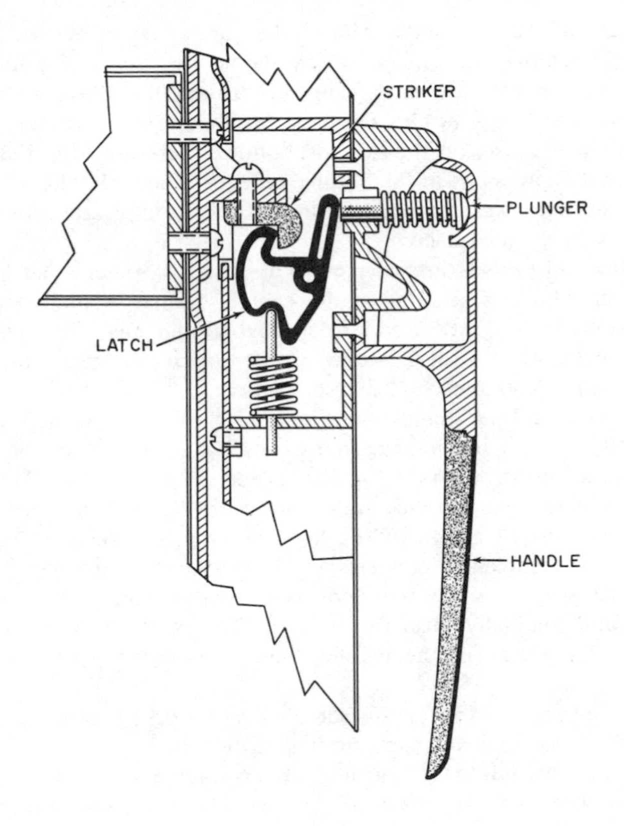

Fig. 6-6. A typical door closure assembly.

Of course, the machine might have to be moved away from the wall, and we should consider how that can best be done. It is fairly easy to move a refrigerator around the room if you slide it on something. Linoleum installers use a felt blanket, but any scatter rug will do. You will have to push at a low point to avoid tipping it over. The simplest procedure is to throw a loop of clothesline over the refrigerator and pull on the bottom.

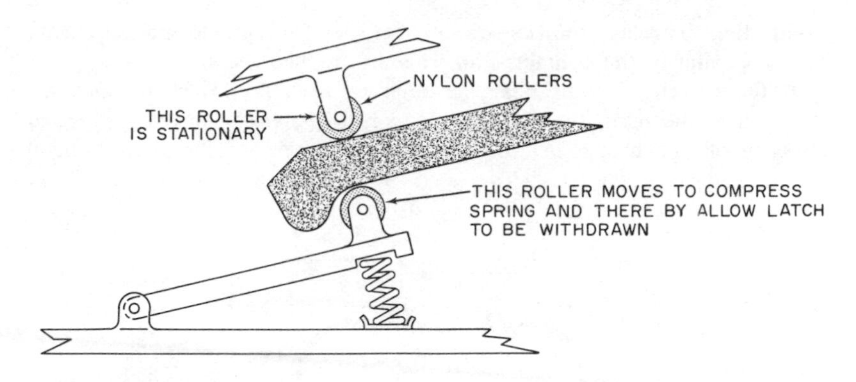
NYLON ROLLERS
THIS ROLLER
IS STATIONARY
THIS ROLLER MOVES TO COMPRESS
SPRING AND THERE BY ALLOW LATCH
TO BE WITHDRAWN

Fig. 6-7. A typical "up-and-down" door closure assembly.

the refrigerator relay

Relating the refrigerator mechanism to the material in the foregoing chapters, and following the same general order of items, we can say that the refrigerator has the same series-wired switch, relay, and three-wire compressor. In the refrigerator, the relay contains the motor protector, and the terminals that connect to the power-supply-cord wires. From the terminals, the usual three wires lead to the compressor and three or four lead to the cabinet light switch and the thermoswitch. There would be four wires, two for each switch, unless there is a common ground wire, in which case there would be three. These are the basic wires, and any others would be for a special fitting such as a butter warmer. Here again, you should be able to find a schematic drawing to show you the function of the wires. As with any household-type equipment, you can duplicate the relay action and start the motor with a three-wire test cord.

In Fig. 6-8 we have one type of relay. It is a "thermally operated starting and overload relay with complete compensation for room temperature." The engineering of such a device is a bit involved, but, simply stated, the principle of operation is this: the current from the switch enters through terminal "L" and goes to the running winding through terminal "M". It also goes to the starting winding through terminal "S". The passage of current through the bimetal starting arm causes it to swing to the right, thereby separating the contact points. The heater element supplies enough heat to keep the starting arm in the "contacts-open" position. On the right hand side, we have the bimetal motor-overload arm. This arm is made so that it will not bend as readily as the starting arm, but, if the motor draws an excessive amount of current, it too will bend, thereby opening the contacts at terminal "L" and disconnecting the relay from its source of power. The toggle springs ensure rapid opening of the contacts and, thereby, reduce arcing.

If the current draw is heavy enough to open the overload contacts, it will heat the reset bimetal enough to bend it, which has the effect of shutting off the heater. At the same time, the tip of the reset bimetal

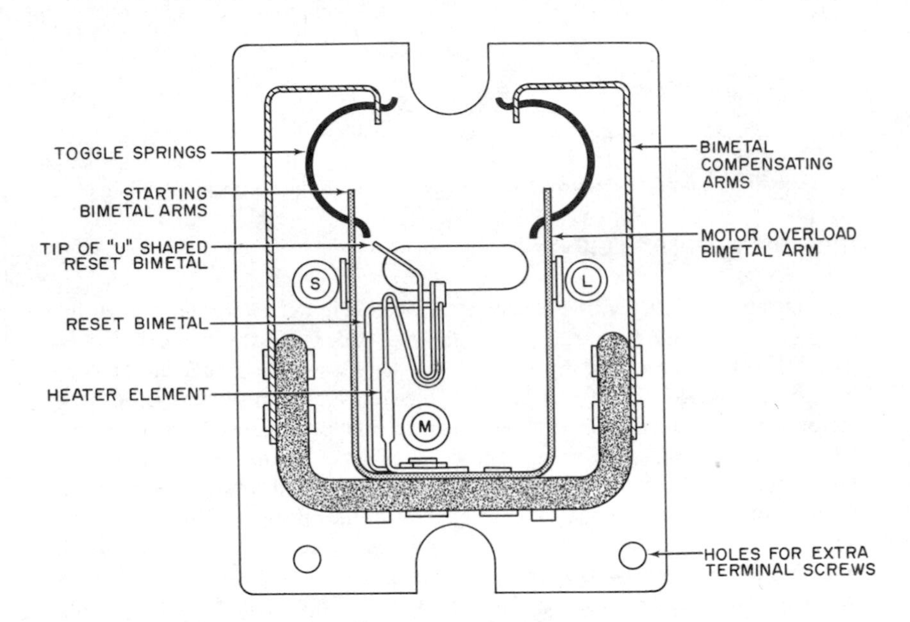

Fig. 6-8. Thermal relay assembly.

will move to the left and touch the starting arm so that it will more readily return to the "closed-contact" position.

The overload arm is made so that it will not cool as quickly as the starting arm, so that the starting contacts close first. With both sets of contacts closed, the relay is ready to start the motor again.

The bimetal compensating arms will bend and vary the position of the toggle springs so that the operation of the relay will not be affected by changes in room temperature. Without the toggle springs, it is obvious that a device of this sort would be unreliable if the room temperature was very low or very high.

Although the relay might have nine or more wires, it can be replaced, without confusion, if you change the wires on one terminal at a time. Some of the manufacturers have used several different relays on a particular size of compressor, but some standardization is provided by requiring only one relay as a replacement for all of them. In this event, the connections might not be exactly the same, and, hopefully, a drawing will be furnished with the new relay to show what wiring changes to make.

the operating cycle

The symptom of a leaking cold-control bellows is rather obvious: alternate defrosting and icing. This can be distinguished from a partly discharged condition by the fact that the defrosting will take place all over the freezer, rather than in one part. In addition, the unit can be timed to give an indication as to whether the cycle is longer than normal. Your wife probably has no idea how long the normal cycle is. Even so, you can deduce something. Of course, how it should cycle depends on the climate and usage conditions.

The normal cycle with an ambient air temperature of 70° F. would depend on the characteristics of the particular cold control being used, but if the unit's cycle was four minutes on and six minutes off, that probably would be normal. On the other hand, if it ran for half an hour and was off for half an hour, there would be little question that the cold control was faulty, except perhaps on deluxe models, in which a long cycle might be normal.

replacing a cold control

The cold control is behind the dial and can be reached by removing the ornamental fixture covering it. The capillary tube will extend through the insulation space and lead into the food compartment and connect to the freezer. The tube might extend for quite a distance between the insulation and the cabinet liner; it might even go around a corner. When you pull the tube out, you would do well to tie a heavy string to the other end so as to be able to pull it back frough the insulation.

What was said before (of air conditioners) about low voltage and starting troubles also applies to refrigerators. Using a three-wire test cord, a 1/7- or 7/32-hp compressor should start without blowing a 7½-ampere fuse.

The refrigerator will not have a running capacitor, but you might find a starting capacitor. You can learn something about the condition of the capacitor by removing the terminal wires, briefly energizing it with the test cord, and then holding a jumper wire on one terminal and touching it to the other. This procedure should result in an arc. When testing small capacitors in this fashion, the normal arc will be so small that you can just hear the "pop" above the surrounding noise level. In other words, the test would not be very conclusive. On a relatively large capacitor, however, if it popped like a firecracker, you could rest assured that the capacitor was in good condition.

If a unit does not start well, you can insert a starting capacitor. The dealer can advise you as to what farad rating should be used, but, if you do not have the factory recommendation for refrigerator motors of 1/7 to 1/8 hp, starting capacitors in the 105-145 microfarad range have been used satisfactorily. Even if the motor will start, it might be well to write the factory and find out what size is recommended. You still might not be getting maximum torque, and, as a consequence, hard starting would be encountered when, for instance, the voltage drops off a little.

In a matter of poor cooling, a possible cause is that normal wear in the mechanical action of the cold control has had the effect of changing the temperature range of the switch. This is nothing to be concerned about; just turn the dial to a colder setting. If the refrigerator runs all all the time but will not cool, however, it is worthwhile to consider whether some household items or newspapers have fallen down in back from the top of the refrigerator. Newspapers in particular can drop down and not be noticed, inhibiting air flow to the mechanism, which will result in an increased cabinet temperature. Some refrigerator units on the market have fans for condensers. Here again, the fan might have stopped, or the condenser might be covered with dust.

It is normal for the warm liquid refrigerant to defrost the evaporator in a very small area where the restrictor is connected. The size of the spot depends on the model. If the unit starts to discharge, however, defrosting is noticed at the point where the heat-laden vapor is drawn into the suction line, that is, where the larger tube is connected. The refrigerant can absorb only so much heat, and when it is fully heat laden, no cooling of the tubes can take place in the area through which the hot refrigerant must flow on its return to the compressor.

overcharging

The volume of refrigerant governs what portion of the freezer is defrosted. The "defrost area" starts where the vapor leaves the evaporator and grows progressively larger as the Freon leaks out. After a while, the defrosted area extends to the restrictor. The discharging takes place over a period of weeks, unless you have a terrific leak, in which case you can hear the hissing, if the general noise level is not quite high.

An overcharge will show up by water dripping on the floor caused by the melting of the frost that forms on the suction line. You will know if the refrigerator is overcharged because the suction line is normally warm, not cold. "Normal" implies starting and stopping (cycling) without the unit doing a great deal more cooling than it would if its contents were already cold and with the door not being opened many times. When first installed and started up, when starting after defrosting, or after having considerable hot food put into it, the suction line can get cold without indicating an overcharge. The condition will remedy itself when the inside temperature drops and the unit runs on a short cycle again.

If the symptoms indicate that there might be a restriction, the indicated procedure is to try to locate it. (A restriction does not have to be in the restrictor). If the restriction is not completely blocking the flow of refrigerant, there should be some frosting in the general area of the restriction. But, sometimes, the restriction just cannot be located, resulting in replacement of some or all of the sealed unit parts.

a faulty defroster

Another cause of insufficient cooling derives from the use of electric defrost heaters. If you find "hot spots" trace the passage of refrigerant through the freezer and decide whether the condition is caused by loss of refrigerant. If not, determine whether a defrost heater is being used on that model. If an electric heater (or heaters) is being energized 24 hours a day, the timer as at fault. Simply disconnect the wires to shut off the heater.

Another type of defrost "heater" uses a switch, somewhat similar to that on the heat pump, to shift the flow of hot refrigerant to a separate system of tubes, which, in this case, are arranged to heat the freezer (see Fig. 6-9). A timer jammed in the defrost position or leakage through the valve will result in a high cabinet temperature.

If the exhaustion of other possibilities leads to the presumption of a discharged condition, locate the leak and repair it. You might have to put in additional Freon to create enough pressure so the gas flow from the leak is noticeable. Nearly all compressors have a "charging" valve or stud that can be used for recharging. The charging stud is a tube on the compressor similar to the one in Fig. 6-10. Notice the fitting that screws into the end. The fitting has slots drilled in it so that Freon can flow in or out of the compressor when the fitting is turned and moved slightly away from the stud.

adding refrigerant

To put in refrigerant, you must use a charging tool see Fig. 6-10. The tool has a ¼-inch flare fitting near its center. The tool is threaded at

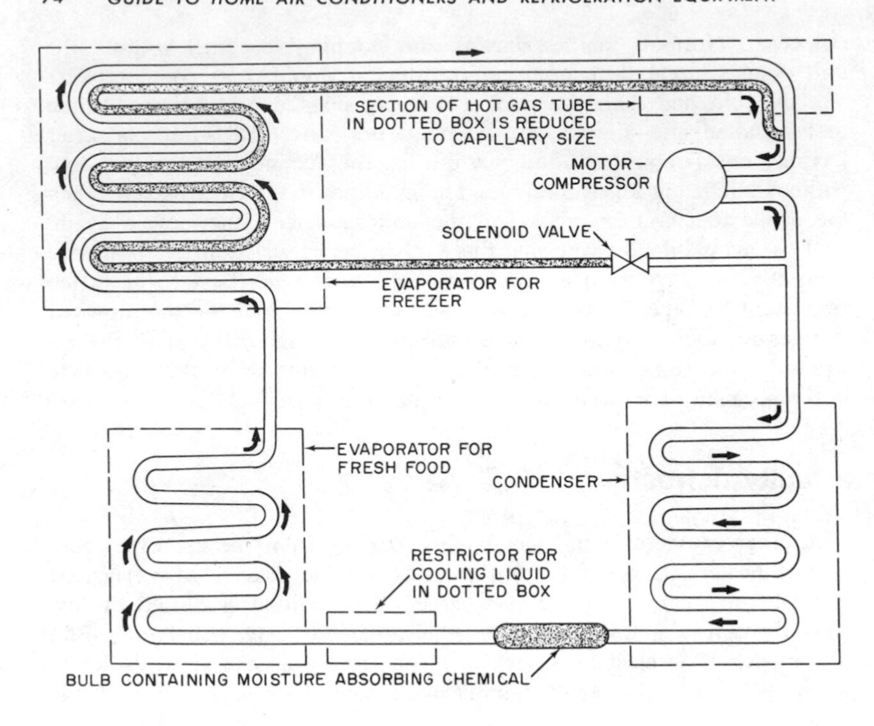

Fig. 6-9. Hot gas tubing for defrosting freezer.

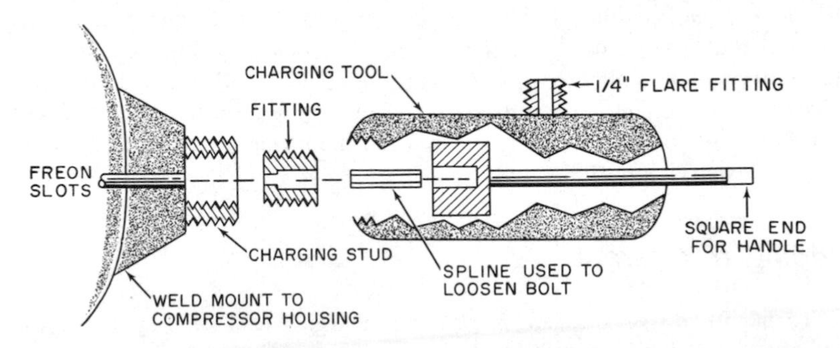

Fig. 6-10. Charging stud and charging tool.

one end so that it can be screwed onto the charging stud, and the other end has packing with a valve stem passing through it. The internal end of the valve stem has a cup that fits over the bolt head. When the external end of the stem is turned, the internal end turns the charging-stud fitting and opens a slot into the compressor. The Imperial Brass Company of Chicago sells a kit consisting of a charging tool and assorted adaptors that will fit almost any charging stud.

Assuming that you can open the charging stud, what then? The unit should have a little plate that specifies the refrigerant and the amount to put in the unit. Do not change the refrigerant as the different types do not mix particularly well. Freon-12 is a standard refrigerant packaged in 15-ounce cans that are available in some hardware stores. A screw-on or clamp-on puncture-type opening that has a ¼" flare fitting is needed. Other gases can be had, but in larger containers. Some are expensive because of high shipping charges.

Assuming that you use Freon-12 and you have everything set up for charging, including either a ¼" copper or flexible tube between the can and charging tool connections, you must remove the air in the tool and the tube in order to avoid the introduction of moisture into the system. With the tube connected loosely to the tool, crack open the charging-stud fittings, and hope there is enough pressure in the compressor to blow out the air in the tool through the loose connection. When you think the air is removed, tighten the connection.

The charging valve is likely to be on the high pressure side. It will be apparent at once if there is any amount of refrigerant left in the system when the Freon does not readily flow out of the can into the charging valve; vaporizing Freon will make the can cold. Do not heat the can by flame or hot water to increase its rate of flow; the can may explode, and, if liquid Freon hits you in the eye, you may be permanently blinded.

When the Freon is going into the compressor, feel the evaporator; when it starts to get cold, close the valve.

Sometimes you can pinpoint a leak by the oil that has run out. If not, you have to search for it. The search might lead to the area behind the plastic molding where the tube goes from the bottom to the top. Some moldings are attached with screws; if not, you might have a problem on your hands. Without the use of screws the molding is fastened by spring clips or flexible "dog ears." The trouble arises from the fact that you do not know which of these two methods has been used. Begin by using a table knife to break the seal between the molding and the cabinet on one side, and the cabinet liner on the other. Then exert pressure to slide the molding toward the food compartment. If it is held with spring clips, they can be bent somewhat by pulling on the edge of the molding (see Fig. 6-11).

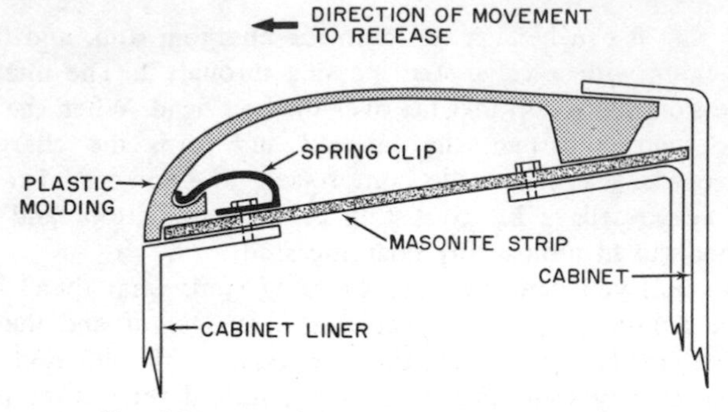

Fig. 6-11. A typical plastic molding placement.

If the leak is where the tube runs through, or adjacent to, insulation, a stethoscope will be of great value. The poor circulation of air in the insulation "segregates" the escaped Freon, with the result that a flame tester will reveal the area of the leak without pinpointing it. Repair is the next step.

soldering

Charging stud and terminal leaks are repaired mechanically by tightening. Elsewhere, it is necessary to fill the hole with solder. Refrigerator soldering is difficult because it frequently involves soldering the bottom of a horizontal tube or soldering on a vertical surface.

Leaks are often found in the evaporator. The evaporator is made of aluminum, which easily can be warped by heat. Furthermore, ordinary solder does not bond with aluminum particularly well. A special low-melting-point aluminum-soldering kit is marketed by the Marquette Company of Minneapolis and is available in automotive supply stores. It contains two different rolls of solder and bottles of flux. One solder roll is actually a brazing rod that melts at 1050° F. To use it on thin-sheet aluminum you need a special torch and some experience. The other solder melts at 400° F. and is ideally suited for this purpose.

To solder anything, it must have a clean surface. In a freezer, you will have some oil to contend with, in addition to a film of food vapor. The area surrounding the leak can be cleaned with carbon tetrachloride; but be careful: this solvent is highly toxic.

A freezer presents a large metal area that continuously draws away the heat applied to the point you are soldering. The metal should be hot enough to melt the solder upon application, but this is difficult to attain because, in using any amount of heat, you run the risk of warping or even melting the sheet metal. In addition, the heat carbonizes the oil inside and might cause a chemical breakdown of the Freon itself. Also, you have to use flux, and if some of it gets through the hole, it will start reaction of its own.

One procedure uses both a propane torch and a "pencil" type soldering iron. (You must clean any corrosion off the "pencil" point before you use it.) The propane torch is used to heat the leak area to decrease the heat dissipated by the metal. After heating the point of leakage, apply the flux. (If the flux changes color, the metal is too hot; allow the metal to cool). Then place the end of the solder strip at the hole and use the pencil point to melt a drop of it, pushing it into the hole. A pencil-point soldering iron works quite nicely for this because the ordinary freezer leak is only a pin hole.

sealing with cement

Another method of sealing is to use epoxolytic cement. A tube of it can be bought at a hardware store and used according to the direction. Epoxolytic cement is simple to use and has been used successfully, but whether epoxolytic sealing will supplant soldering for evaporator leaks is a question that stands unresolved so far.

Because the freezer is on the low, or suction, side of the system, it follows that some air might have been drawn in through the leak. In theory, you should use a vacuum pump to remove all the air and Freon (called evacuating) and recharge the unit every time a leak is discovered. The evacuating procedure will be described in Chapter 8.

When determining whether it is worthwhile to search for a leak, it is helpful to know the running time of the unit. When the refrigerant charge decreases, the running time increases. In other words, if it takes one hour of running time to complete the cycle with a certain amount of Freon, it will take two hours to attain the same cooling effect with half as much. Maybe it does not work exactly in that manner, but that is the general idea. The efficiency of the system is only slightly decreased because the pump would have decreased low-and-high pressures to contend with in the presence of less refrigerant.

checking the cycle

Modern refrigerators run so quietly that most of the time the housewife simply would not know how long it was running. We know that a sticky-bearing-and-poor-relay situation might prevent it from operating on a normal cycle, even if it appeared all right when listening to it. And, in terms of the cooling that the unit is required to do, it might be that the cooling capacity is exceeded because of high room temperature, a leaking seal, or a dirty condenser.

If, in the analysis, the running time of the refrigerator is determined, you can use that as a stepping stone to further deductions. One way to do this is to take an electric clock and wire it to run when the compressor does. The clock can be attached to the white terminal in the relay and to the running winding terminal.

If the unit runs most of the time and does not cool, it may be presumed that, if the unit starts and stops occasionally, the switch and relay are all right. If the condenser is very hot, there is probably air in the system.

If the unit runs all the time and does not cool, which could very well happen in a hot climate or in summer, the answer might be simply that an excessive number of beverage bottles or cans were being chilled. Chilling warm liquid will make the unit run more than anything else.

volt-wattmeter readings

Although the foregoing procedures may be used to arrive at a deduction, they do not provide a means of quickly getting a direct indication of what is wrong. A volt-wattmeter can be used for that, but even this is not 100% foolproof. You can have a combination of things wrong that will throw the readings off. For instance, you can have a normal wattage reading with sticky bearings and loss of refrigerant.

With the volt-wattmeter shown in Fig. 6-12, one can read watts on the left and volts on the right. This model has three wattage ranges and two for volts, while some have only two wattage ranges.

Fig. 6-12. A volt-wattmeter. (Courtesy of Robinair.)

TABLE 6-1: CHARACTERISTICS OF SOME 1/8 HP MOTORS

Characteristic	Range With Capacitor			Range Without Capacitor		
	100v	115v	125v	100v	115v	125v
Running wattage at						
70-90° F	105-128	108-138	114-144	105-128	108-138	114-144
90-110° F	125-155	133-163	139-169	125-155	133-163	139-169
Starting wattage						
Normal stall	900-1000	1250-1350	1525-1625	1700-1800	2200-2350	2500-2700
Stalled Shorted Capacitor	1700-1800	2200-2350	2500-2700			
Open started winding	500-550	700-750	850-900	500-550	700-750	850-900

Power can be computed by taking the square of the voltage divided by the resistance, among other ways. Rewriting this expression using Ohm's law (E=IR), will reveal that voltage has some connection with the wattage reading. The voltmeter will show how low the voltage drops at the time of high inrush when the unit is starting.

When using a volt-wattmeter, set it on high range for starting and switch to low range after the motor has speeded up. Several things can be determined with this device. With a normal start, the needle will "overshoot," and travel beyond the reading for the starting and running windings combined, but it will quickly drop back to where it should be, and, when the starting winding disconnects, it will drop to the running-winding reading.

With a normal start, the running winding will energize quickly, and the needle will "dwell" only momentarily on the starting reading. Thus, the meter will tell you the time required to energize the starting winding. In other words, it tells what the relay is doing.

With a cold motor, the reading will drop somewhat as the motor warms up. You will not learn very much when the unit starts. You have to wait until the refrigerant circulates to establish a normal relationship between the high- and low-side pressures. How long that will take depends on circumstances, but it might be nearer 20 minutes than 5.

Because the pressure of the refrigerant is related to air temperature and the capability of the condenser to give off heat, and because consumption of power by the motor is related to the voltage received, you will get a different "normal" reading for every change of air temperature or voltage. With a given normal reading for a particular temperature and voltage, the reading will be 15% low with a shortage of refrigerant, 35% low with a restriction, 50% low with the valve jammed open in the compressor, and 15 to 20% high with air in the system or with a dirty condenser. If the starting winding does not cut in, the reading will be high; if it does not cut out, the reading will be very high; and, if badly shorted, it will be "sky" high. Table 6-1, which is for a particular ⅛-hp motor, shows the scale or relationship of the readings with various situations.

7—Deluxe Model Refrigerators

odors

We can begin by discussing a problem that is encountered with conventional as well as deluxe models—that is an unpleasant odor. Food compartments generally have an odor associated with them, even though cleaned frequently. It might result from putrification of a food (such as gravy) that had run down into the space between the cabinet liner and the molding where it could not be washed out. Sometimes the odor is quite bothersome.

The manufacturers are aware of this problem. They do not only hear about it from service managers, but they receive letters from irate housewives. One manufacturer made a study of the matter and concluded: "Since the cabinet temperature is the same in the summer as it is in the winter, these summer odor complaints must be a psychological matter or a chemical phenomena that defies explanation."

The author has his own notion as to the reason. The odor is the same the year around, but the kitchen air-circulation pattern is not. In the winter, the heating plant keeps the air moving, and foul air is "diluted" by other air as it moves throughout the house. In hot, damp weather, the air does not circulate, and the odor "accumulates," as it were.

That, however, does not answer the question of what produces the odor. Minor causes include an electric heater that moves slightly and touches the plastic molding. Major causes include ice cube trays and the plastic molding that acquire a charge of static electricity, making them attract food vapors of a sort that readily produces a bacterial action and give off an odor. A certain polish will neutralize the static and remedy the situation. The fancy term for it is "Silicone Anti-Static Plastic Polishing Compound."

the refrigerator-freezer

To return to the subject of refrigerator-freezer models, one of the early practices was to install a separate door on the across-the-top freezer. That was an excellent arrangement because it prevented warm air from entering the freezer every time the food compartment door was opened. In addition, frosting was reduced (frost is only condensed air moisture), and less moisture-laden air was being introduced. The extent of frost trouble in a freezer depends on use, that is, it seems as if the so-called "deep freeze" (where the *entire* space is a freezer) is opened less than a small freezer (usually combined with a refrigerator).

The refrigerant circuit pictured in Fig. 7-1 has several interesting features. The bottom layer of freezer coils has fewer passes because cold has a tendency to settle, requiring less cooling area at the bottom.

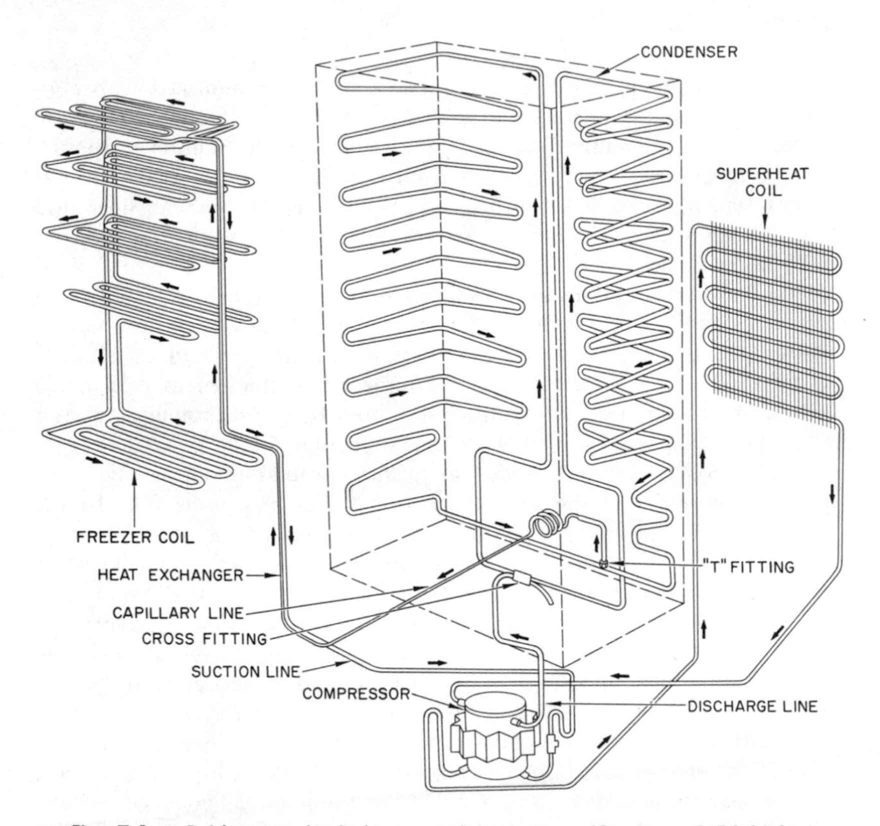

Fig. 7-1. Refrigerant circuit in an upright freezer. (Courtesy of Frigidaire.)

The arrows show that the warm vapor is pushed into the superheat coil where a large part of the heat is removed. The warm vapor then goes into condenser coils in the refrigerator walls. The wall coils are in right and left sections; Freon is directed into each section. The liquification of the refrigerant is completed in the wall coils, and, at the "T" fitting, the Freon discharged from the two sections merges and enters the capillary line to complete the circuit.

Refrigerator-freezer combinations that have little or no frost have been marketed in the past decade or so, and their operational principles are somewhat more complex. Complete diagnostic procedures for all the models that the various companies have marketed would require a huge book; we must content ourselves with some generalities.

the electric defroster

After World War II, some units were made with a device that simply shut off the compressor for a hour or so at night. This arrangement allowed the frost to melt, but it also hastened the spoilage of food. The public felt that more sophisticated methods were in order. In recent units, frost removal is accomplished by one of three general methods: the use of electric heaters or, perhaps, hot-gas coils; the use of a very long cycle with the frost melting during the "off" period and running down a pipe to a pan in the bottom where it is evaporated; or the second method combined with "sublimation," which will be explained later.

Because electrical controls have moving parts, they are subject to mechanical difficulties. For example, a control might work perfectly for 99 times and then "stick" either open or shut; it might release itself and then return to normal operation for an unpredictable length of time. With sophisticated systems, the interrelation of the effect of a partial discharge of the sealed unit combined with erratic operation of the controls presents a diagnostic problem that is not as simple as it is with a "basic" system. Because the purpose is to melt the frost quickly and hold the thawing of food to a minimum, the electric heater has to be of fair size with, say one-third the heating capacity of a flat iron. This is all well and good when the unit is fully charged, but, unless there is a specified amount of Freon to carry away the heat, the evaporator will heat up terrifically. If the evaporator becomes too hot, a thermally actuated device will shut off the power supply, thereby compounding matters.

In this case, if the heater was cold when you checked it, it would be hard to say whether the heater had stayed on too long or whether a partially discharged condition was present. No doubt, a wattmetter reading would indicate if there were a shortage of refrigerant, but you might be able to resolve the question without a wattmeter. Use the schematic to determine which wires go to the heater, and disconnect them. Leave the unit alone for a few days until the evaporator frosts up. Then, by studying the frost pattern, you can decide if the evaporator has discharged refrigerant.

Some electric heaters are timed to go on after a specified number ot hours of compressor operation. Thus, if it defrosts in the daytime, it does not necessarily mean that the timer is out of order.

the hot-gas defroster

Figure 6-9 shows an evaporator tube that carries hot gas from the condenser at specified times. The hot-gas tube might not be as long as the cooling tube because the heating coils work more effectively, and not as much radiating area is required. In other words, the hot-gas tube might take "shirt cuts" and not follow the cooling coil exactly.

For defrosting, the timer energizes a solenoid that raises its plunger, opening an orifice and allowing the hot gas to flow into the other tube. If the unit will not stop defrosting, it would be worthwhile to apply the test bulb wires to the solenoid terminals. If the bulb lights, indicating that the solenoid is energized, it appears that the timer is out of order—perhaps the clock stopped when the contacts to energize the solenoid were closed. If the bulb does not light, this indicates conventional troubles or that the hot gas was still going through the de-energized valve. This, in turn, means that sticky oil is binding the plunger or that the spring is broken. If the plunger is stuck, gentle tapping might free it.

Opening the passage into the defrosting coils reduces the high-side pressure and reduces the flow of Freon into the cooling coils. Because a little extra Freon is needed when both coils are in use, an "accumulator bulb" serves as a "storage place" to hold the extra Freon at times when the unit is not defrosting.

The other end of the hot-gas (defrost) coil connects to the suction line. When an increase in cabinet temperature is noted, its cause might be any of the conventional refrigerator malfunctions, or it might be hot gas leaging through the solenoid valve.

Another method of keeping the unit from frosting uses an evaporator plate that collects frost when the unit is running, but when the unit shuts off, the frost melts and drips into a pan located where the heat of some loops or "passes" of the condenser tube will evaporate it. In Fig. 7-2 we see water dripping from the plate into a plastic tube and running down to a pan. The same thing occurs in the bottom (freezer) compartment. When the power goes off, the frost melts and drips down to the pan as shown.

The temperature differential between the freezer and the refrigerator sections of a combination unit varies somewhat according to make, model, and operating conditions, but it might be as much as 37°F. Thus, the temperatures might be 40° F. in the refrigerator and 3° F. in the freezer. If you check the temperature range of the switch, you will

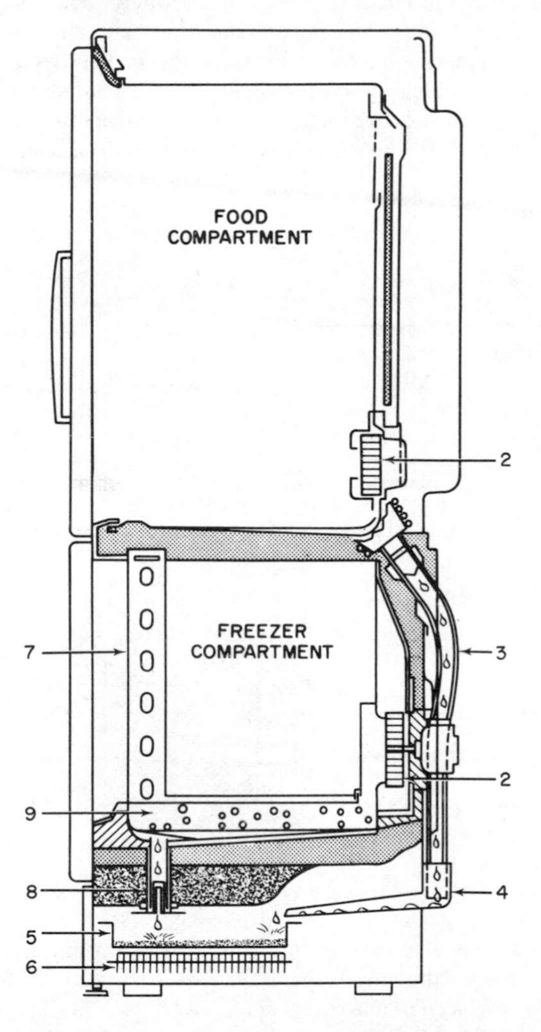

Fig. 7-2. Frost dispostal during "off" part of cycle: (1) evaporator plate, (2) fans, (3) plastic tube, (4) trough, (5) pan, (6) condenser, (7) freezer air distribution duct, (8) freezer drain, (9) chill coils. (Courtesy of Frigidaire.)

find it will indicate a higher temperature than the actual freezer temprature. That does not mean that the switch is not calibrated properly.

Actually the switch bulb, which is bolted to the sheet metal between the tubing outside of the freezer compartment, is affected by whatever heat comes through the insulation. Therefore, the switch is governed by the temperature of the metal to which it is bolted. The coils have to be colder than the "cut-off" point to make sheet metal, with the switch bulb, the same temperature as that found inside the freezer.

To get a pictorial idea of what goes on in a unit of this sort, examine the time-vs.-temperature graph for a well known product (see Fig. 7-3). This situation would be fairly normal with moderate operating

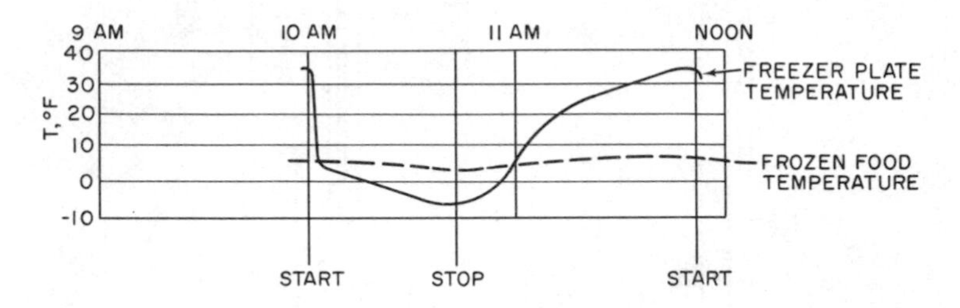

Fig. 7-3. Time vs. temperature for normal operating conditions.

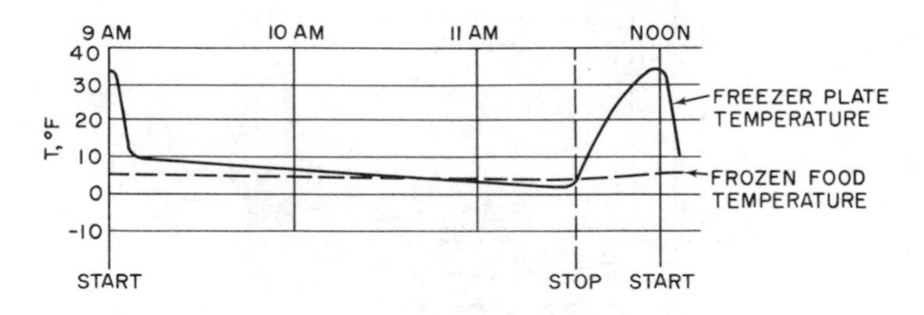

Fig. 7-4. Time vs. temperature for unusual operating conditions.

conditions, but what happens when the unit is operating under abnormal conditions? Suppose that the room is 95°F. and a lot of cold food was removed and replaced with warm food (and beverages). In this case (see Fig. 7-4), after an initial drop of more than 20°, 2½ hours would be required to drop from 10°F. down to zero. After that, the unit (unless more replacement of food took place), would assume a cycle something like the one shown in Fig. 7-3.

Note that, in spite of the considerable variation of the freezer coil temperature, the temperature of the frozen food does not vary much, only a few degrees. Note also that, after a prolonged period of intense cold in the freezer as shown in Fig. 7-4, the cold-control switch shuts off the motor at a slightly higher temperature.

the "remote reading" thermometer

In Fig. 7-5 we have a device used for sensing the temperature changes inside the cabinet while the door remains closed. This "remote reading" thermometer has a cable (stored in the back when not in use) with a sensing probe. With the case resting on top of the cabinet (or wherever is convenient), this "senser" can be placed inside either the frozen or fresh food compartment and the door closed on the cable.

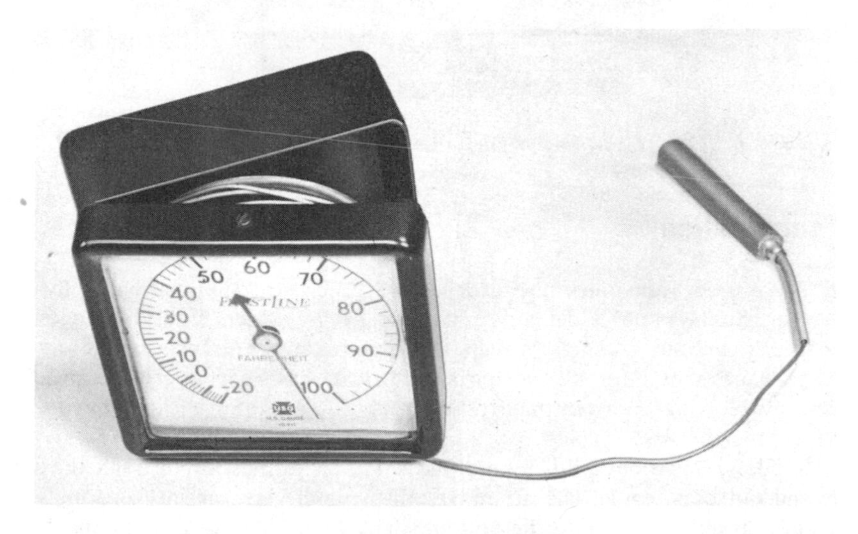

Fig. 7-5. Single range, one probe, remote reading thermometer.

Figure 7-6 shows a more elaborate device of this type. It has two temperature ranges that permit readings from –50° to 150° F. The cables, or probes, can be used at the same time. All you have to do is turn the selector knob to give a reading on the prob you wish to read. This practice of using three probes is very handy when working on a air conditioner, because the probes can be put in three different places to check the circulation of air.

To return to trouble-shooting procedures, we may say that with a combination unit, a refrigerant shortage or a restriction will cause variations of temperature that can be plotted in a graph such as that shown in Fig. 7-4. The readings for such a graph can be obtained with a remote-reading thermometer and a clock.

Fig. 7-6. Dual range, three-probe remote reading thermometer. (Courtesy of Robinair.)

"sublimation"

These units sometimes make use of the process of sublimation in the freezer. Sublimation is defined as the process by which a solid changes to a gas without passing through the liquid state. This process occurs when a cake of "dry ice" evaporates. In our case, we apply the process by blowing air over the frost so that it evaporates without having to be melted first.

In Fig. 7-7 we have this taking place. Fan E pushes air through duct F and out opening 1. The air moves through the freezer into openings 1 and through duct 3 into the bottom space where it passes over coils 4. Figure 7-8 shows the cage-type fan that is used for air circulation. The fan motors used for this service are self-contained, permanently lubricated motors, and it is believed that they can run continuously for at least ten years before breaking down. Figure 7-9 shows how motor, fan, and duct are mounted on the back of the freezer. The air moves through the duct for across-the-top distribution.

Sometimes small timer-actuated electric heaters are used with these units. Their use would be just a precautionary measure—in case something (such as spilling a pan of water) caused frost or ice formation greater than that which the process of sublimation could cope with.

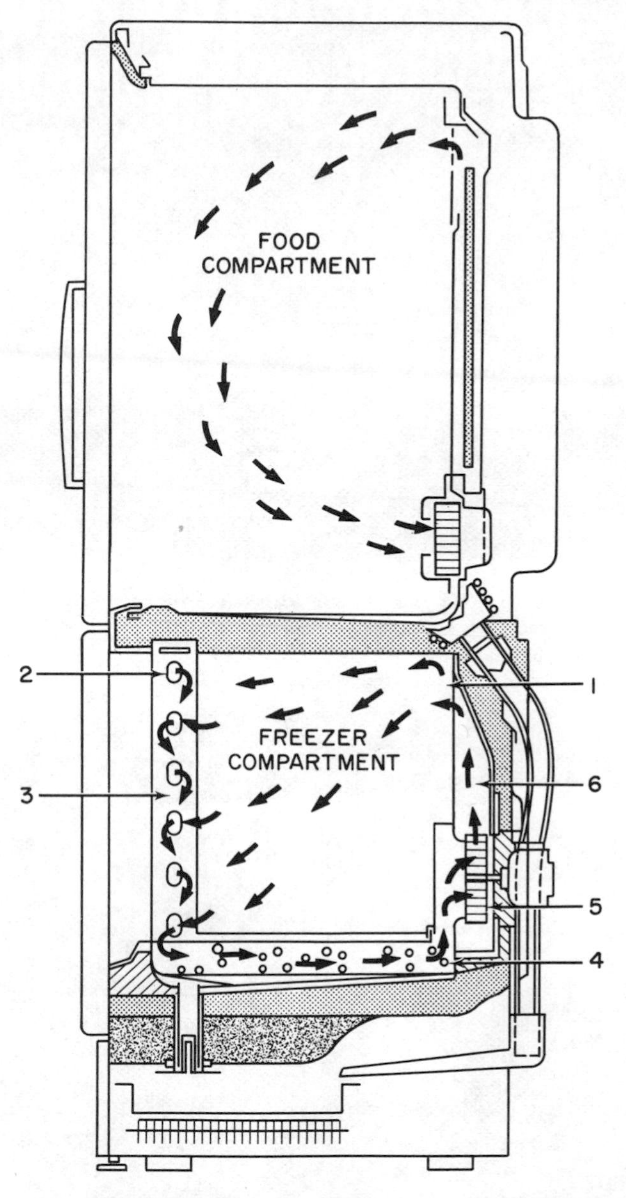

Fig. 7-7. Air circulation pattern: Fan (5) pushes air through duct (6), out opening (1). It moves through the freezer and into openings (2) and through duct (3) into the bottom space, where it passes over finned coils (4). (Courtesy of Frigidaire.)

Fig. 7-8. Cage type fan used for air circulation: (1) condenser, (2) rubber gasket, (3) fan, (4) opening into freezer, (5) duct, (6) gasket holder, (7) compressor. (Courtesy of Frigidaire.)

Fig. 7-9. Fan mounted on rear of freezer: (1) freezer liner, (2) duct, (3) rubber gasket, (4) fan motor. (Courtesy of Frigidaire.)

8—Other Tests and Procedures

The heavier unit has a 208/230-volt or three-wire input. This equipment operates in the same manner as the 115-volt equipment, but the condenser might be water cooled and the restrictor is likely to be replaced by an expansion valve.

the expansion valve

An expansion valve is needed to handle the larger volume of Freon that passes into the evaporator in these units. There are several types of expansion valves, but, on household units, it would probably be of the thermal type, which means, simply, that it has its own gas-filled bulb, tube, and bellows that senses a temperature change and causes a mechanical action, which, in this case, is the opening or closing of the intake port in the valve. In Fig. 8-1, we can see that expansion of gas in the bulb will result in pressure being exerted on the diaphragm, which in turn transmits that pressure to the plunger cap. This compresses the spring, and pushes down the plunger shank, causing the plunger body to uncover the intake port.

On relatively heavy household or light commercial equipment, a solenoid could move the plunger to accomplish the same result. But, with either type, a drop of water falling on the junction of the needle point and the orifice, can freeze, and, thereby, shut off the flow of Freon through the valve.

When a unit with an expansion valve will not cool, it is worthwhile to pour hot water on the valve. This will melt any ice that is present. If that does not help, try tapping the valve. If it has a solenoid, use the test bulb to see if it is energized. If the test fails, the solenoid coil may have a broken wire. You can test a solenoid by taking it off and putting a nail in it and energizing it. If there is a magnetic field, the nail will jump in a most interesting fashion. Of course, the lack of cooling might stem from lack of Freon. Whether or not this is the case can be determined with a pressure reading, which we will discuss later.

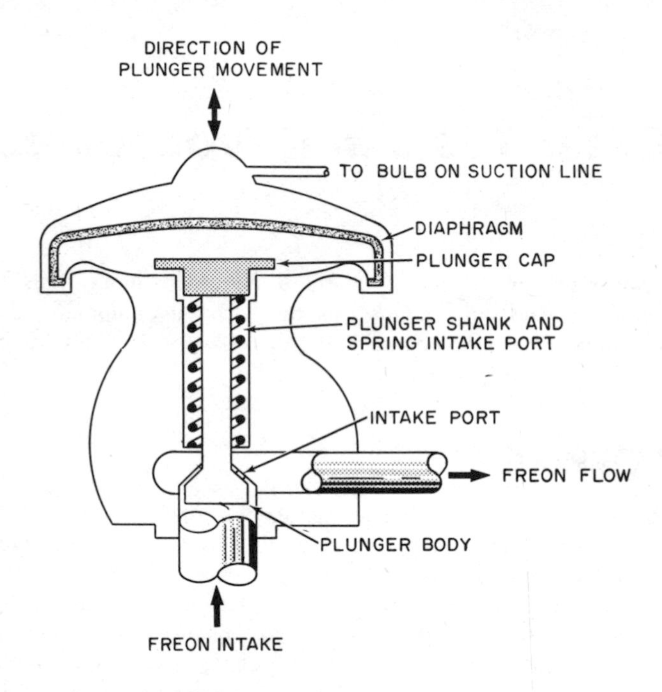

Fig. 8-1. Thermal expansion valve.

remote condensers

As for condensers, it is apparent that, on an air conditioner at least, you can not very well blow the hot air out the window if the unit is in the middle of the house. In this case, the condenser has to give its heat to water, which, in turn, gives its heat to the air. Or, at least, that is the usual arrangement. Now, you could just let the warm water run down the drain, but, in most places, water costs enough so that it can not be wasted. Thus, the cooling water is pumped around in a "closed circuit" with the heat being given off to the air in a type of radiator. This "radiator" is likely to be on the roof, but wherever it is, the water will not flow without being pumped. So, if there is something wrong with the pump, the house will be warm.

A common difficulty with these water-cooled condensers is that a "hard water" condition results in the accumulation of scale on the tubes. The scale is likely to form over a period of years, and the unit slowly loses its efficiency. When scale formation is the problem, quite a bit of the scale can be removed by opening the tank and scraping it with a wire brush. Of course, there are chemical ways of removing scale. The Calgon Company manufactures some preparations that are intended for this service. One is claimed to "anticipate" scale formation and prevent it; the other is to be used after scale formation has already taken place.

In the old days of "open" units, the standard testing procedure was to take pressure readings on fittings provided for this purpose. On a modern sealed unit, however, provisions are not made for pressure readings. The volt-wattmeter is used, and it, combined with the use of the sense of touch (relative to the temperature of the various parts of the system) and hearing, will provide a diagnosis in all but the most complicated cases. Electrical testing should be done first because taking a low-side pressure reading involves cutting into the system (cutting through a tube or punching a hole in it), and that should be done only as a last resort. Of course, the equipment might have a fitting for a low-side reading, but, on household units, it is most unlikely.

compound gauge and manifold

If pressure readings are necessary, a manifold assembly has been devised that has two gauges mounted on it. One gauge, for the high side of the system, reads from zero to, say, a few hundred pounds per square inch (psi). The other gauge, for the low side, gives a "back-pressure" reading somewhere between say, 150 psi and 30 inches of vacuum.

The manifold also has three intake ports (¼ flare) with a valve stem on each end. With the valve at one end open and that at the other end closed, a high-side reading can be taken, and, when the valves are reversed (the open one closed and the closed one open), a low-side reading can be taken. The third intake has a purpose, but, when the manifold is used simply for testing, it can be capped.

One situation in which the manifold will indicate that something has gone wrong inside the compressor is when the pressure on both sides of it is the same. For instance, if the high side, or "head" pressure is 75 psi and the "back" pressure is 75 psi, obviously something is wrong with the compressor.

temperature vs. pressure

In other situations, the readings can give "presumptive evidence" of a malfunction. To gain an insight into how it works, let us consider how the figures on a refrigerant temperature-vs.-pressure graph are obtained. Let us say that you have a flask of Freon-11 in a room in which the temperature is exactly the same as the vaporization temperature of the Freon (in this case, 75° F.). If the flask is full and you cannect a compound gauge to it, you can vary the temperature and get the same readings as on the graph. That is, if the room was heated, the Freon would expand and increase the pressure in the flask. How much the pressure increased would depend on how much the temperature increased. At 90° F. it would increase 5 psi and at 100° F. it would increase 8.9 psi. At 0° F. it would decrease to 24.7 inches of vacuum and at –20° F. it would decrease to 27 inches of vacuum.

Of course, it is unlikely that you would find Freon-11 in household equipment, but we are using it as an example to show that with a room temperature of 65° F., the normal readings would not be the same as with the room at 95° F. In addition, a shortage of refrigerant would make a difference. If there is no other way to reach a conclusion, the unit can be discharged of the refrigerant, and then it can be recharged with the amount specified on the unit information plate. You must take care that air does not enter into the system.

We might see the usual high- and low-pressure relationships by relating some sample readings to a particular refrigerant and room temperature. With Freon-12 in a room at 90° F., the head (high-side) pressure should be 100 psi or higher, and the back (low-side) pressure should be 8-12 psi. If the head pressure is 100 psi and the suction side has 25 inches of vacuum, the unit is restricted. With a high suction reading and rather low head pressure, it would be worthwhile to add some Freon and see if normal readings will result.

pressure readings

Taking a head-pressure reading is simple—all you have to do is screw the charging tool onto the charging stud (using the proper copper gasket) and attach a pressure gauge directly to the tool with a piece of tubing, which could be ¼-inch copper tubing or one of the flexible tubes with knurled "finger-tight" tightening attachments on the end. At least, that is what you can do if the compressor is fitted with a charging stud.

It is not unusual to encounter a compressor that was charged through a "process tube," which is just a length of tubing extending out of the compressor housing. The person doing the charging just crimped or flattened the tube, cut it off, and soldered it when the charging was completed. To get a reading from a compressor with a process tube, you have to use a fitting the same as you would use to take a back pressure reading. Now we will see how this can be accomplished.

If you decide to use the following ideas, take warning: by punching holes in, and otherwise tampering with, a sealed unit, you run the risk that moisture will get in and carbonize the oil. This would not necessarily happen right away; a leak will eventually cause the suction side to develop a vacuum, which, in the case of a low-side leak, will cause air (moisture laden, more or less) to be drawn in. The eventual outcome will be a stuck compressor or a serious restriction. If the unit is in warranty, puncturing or cutting the tubing will void the warranty.

The preferred way to take a suction-line reading is to install a "T" block that has three flared openings (see Fig. 8-2). It can be attached

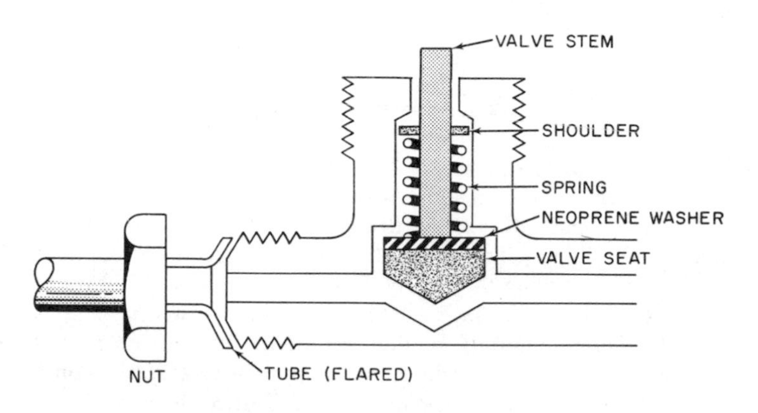

Fig. 8-2. "T" block fitting.

to the (cut) tubing with flare nuts. The part that extends from it at a 90-degree angle has a valve. Taking a reading requires a special attachment that will depress the valve stem and open a passage into the tube. When the attachment is removed, the pressure of the gas, assisted by a spring, pushes a valve seat (plate) against a neoprene washer to close the opening. When one of these "T" blocks is installed by a repair shop and then properly capped after being used, the chance of leakage should be reduced substantially.

A suction-line reading also can be taken with a device that clamps on to the tubing and has a neoprene block that bears against the tubing (see Fig. 8-3). The neoprene has a hole in the center through which

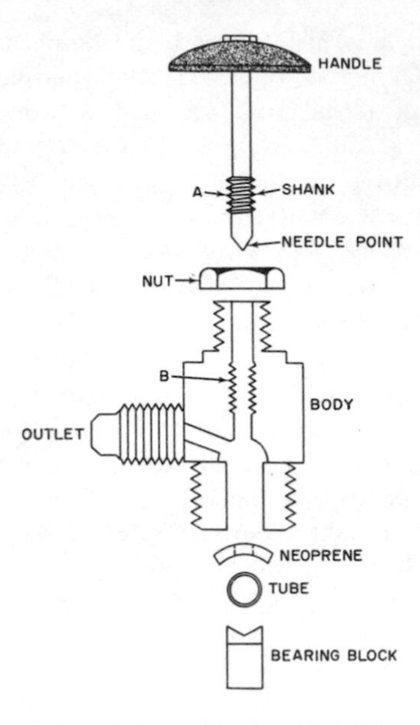

Fig. 8-3. "Clamp-on" fitting.

a pointed shaft passes. Turning a handle on the end of the pointed shaft moves the point away from or toward the tubing. After being clamped on, the device is put into use by turning the handle to cause the point to puncturing the tubing. (The gauge should be connected, of course.) With a passage into the tubing thus opened, a reading can be taken by backing the point out of the passage. After the reading, the passage is then closed by turning the shaft so that the point fills the hole. After the flexible hose or tube to the gauge has been removed, putting on a flare cap will reduce the chance of leakage. With respect to leaks, this fitting is less satisfactory than the "T" block.

replacing a sealed unit

If you needed to replace a sealed unit, and wished to make the replacement yourself, this would be the procedure. (We will use a refrigerator for an example.) First, decide whether it is to be installed from the front or the back. If the cabinet back is solid, the evaporator has to go out through the door opening. If the suction and liquid lines run up the back, and if the evaporator has a plate somewhat larger than itself in back of it that bolted to the cabinet, removal from the rear is indicated.

Second, detach the cold-control bulb from the evaporator and the wires from the compressor. Then expose the tubing. If the tubing is on the back, you just remove a piece of sheet metal. On the front, it is necessary to remove one or more of the plastic molding strips, plus some sheet metal pieces, and, if the molding has masonite behind it, that has to come off too.

(What is being described is the procedure for a conventional refrigerator in which the evaporator coils are in the freezer at the top. If the coils were in the walls, the cabinet liner would have to be unfastened and pulled out.)

Third, unbolt the motor-compressor and whatever braces that would prevent it from being pulled out of the cabinet. When the freezer mounting bolts are removed, the unit will be ready to be removed. The difficulty at this point will be that you will not have enough hands. Getting the compressor out will be expedited by using some pieces of wood to slide it on, but in moving either the compressor or the freezer, and because they have about five feet of tubing between them, you will find a helper useful.

It can be done single handed by getting something about 30 inches high to rest the freezer. The replacement unit comes in a crate that can be used.

Fourth, pull out the freezer and rest it on the crate. (Try not to kink the tubing.) Then you can move the compressor a few inches, slide the crate and freezer over the same distance, and keep moving it until the unit is freed. As a space saving measure, the tubing on the replacement unit would be looped so that the unit could be shipped in a crate smaller than the refrigerator. You will have to re-form the tubing so that it will have its original conformation. By the same token, the tubing on the old unit will have to be bent if it is going to be crated and shipped anywhere.

Fifth, seal the cabinet against air infiltration with some sealing compound such as Permagum. If sealing compound cannot be obtained readily, ordinary caulking compound can be used. This is the kind that comes in a "gun" and is used for filling in cracks around the windows in homes.

Sixth, reconnect the cold-control bulb and the motor wires—the unit is ready for use.

replacing a component

To make a component replacement, the first step is to make sure the tubes are of the same diameter. If not, you will have to get special connectors. Flare connectors are most widely used for this purpose, but there are other kinds that can be used, such as heavy-duty forged connectors.

A component replacement requires a vacuum pump. Specially made ones are available, but any motor-compressor will serve. One can be obtained from a unit that has been scrapped for some other reason. The suction line would have to have a ¼-inch flare fitting soldered to it.

A supply of Freon is also required. A professional would use a charging stand that has a "sight glass" that is marked in ounces of Freon (see Fig. 8-4) for withdrawing the exact amount of refrigerant.

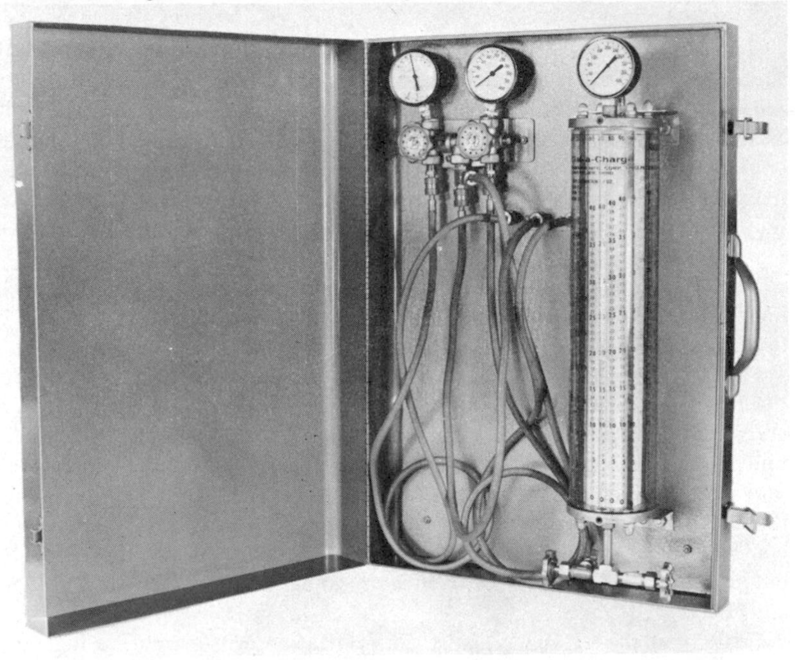

Fig. 8-4. Charging kit with manifold, gauges, connecting hoses, and freon holder.

However, the job can be done without a charging stand. Fifteen-ounce cans can be coupled together with tubing to supply the required amount, whatever it might be. Of course, you have to be careful not to overcharge the unit with this arrangement.

To set up the equipment (see Fig. 8-5), put the charging tool on the charging stud of the compressor (the new one, if it is being replaced), and connect the charging tool to the middle port on the manifold. Connect the vacuum pump to the compound-gauge end of the manifold and the Freon supply to the remaining port.

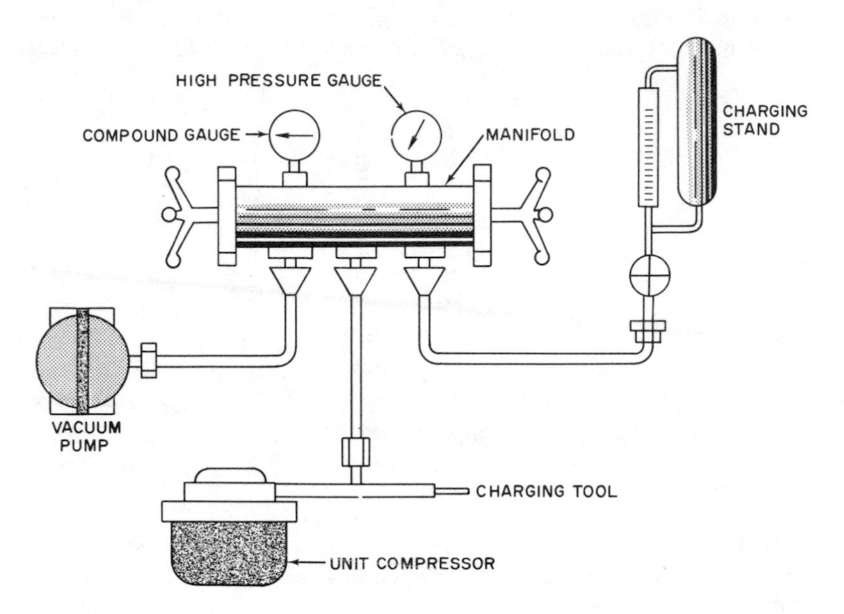

Fig. 8-5. Equipment connections of charging kit.

When you are ready to cut the tubing, cut, couple, and evacuate as quickly as possible so that the interior of the sealed unit will not be exposed to the air any longer than is necessary. With the tube ends coupled together, start the pump. The charging stud on the compressor should be left closed. The idea is to see if the fittings are leaking. Let the pump run for five minutes. The vacuum reading should be about 30 inches. Stop the pump and let everything stand for five minutes. The vacuum reading should be the same at the end of the period. If not, you will have to locate the leak and repair it. One source of leakage is the manifold valves. The stems have to be turned either all the way in or all the way out.

Now, open the charging stud by turning the fitting with the handle on the charging tool, and repeat the same procedure as outlined above. In this case, the evacuation and testing times are longer: Pump out for 20 minutes and hold a vacuum for 20 minutes. A perfect vacuum is difficult to attain, but the reading certainly should not drop below 27 inches. If the vacuum holds, close the valve to the pump, and let in the Freon. After the correct amount of refrigerant is in, close the charging stud and start the unit motor. Let it run for a few hours to determine that it is not over or undercharged.

Figure 8-5 shows how the various connections are made. Of course, it does not *have* to be done this way, but, if you use this procedure, you should have the following state of affairs at each stage of the program:

	Valve on Charging Stand	Vacuum Pump	Valve on Vacuum Pump	Charging Tool	Valve for line to Charging Stand
First Evacuation	Shut	On	Open	Shut	Open
Test Fittings	Shut	Off	Shut	Shut	Open
Second Evacuation	Shut	On	Open	Open	Open
Test for Leaks	Shut	Off	Shut	Open	Open
Charge	Open	Off	Shut	Open	Open

At the end of charging, close all valves and disconnect. Note: Valve for line to charging stand has to be open to evacuate air from the line. If charging stand valve is leaking, obviously the valve where the line connects to the manifold will have to be shut.

When replacing a component part of a sealed unit, you should install a "dryer" on the high-side line. The technical term for it is "conical screened molecular sieve dessicant container." It is just a bulb with a tubing connection at each end, but it has to be installed so that the Freon will move through it in a specified direction. The dryer is filled with granules of a chemical compound. The Freon has to pass around these granules that have the property of attracting and seizing whatever water is mixed with the Freon until they become saturated. At the outlet is a cone-shaped screen. Any solder, dirty oil, etc., that comes along is supposed to be pushed down to the base of the cone and not obstruct the subsequent passage of Freon.

refrigerant properties

At this point it would be worthwhile to briefly mention certain properties of refrigerants. If Freon is subjected to intense heat (or on fire) it will break down chemically and become a poison gas. The two other refrigerants that were popular years ago and still may be encountered are sulphur dioxide and methyl chloride. The former has the properties of tear gas. If any amount of it is in the air, you will need a gas mask.

To find point of leakage soak a rag in ammonia and hold it at suspected leakage points. Direct contact of the two will result in the formation of a heavy cloud of white smoke. As for methyl chloride, people using a flame type leak tester have found that it will burn. Furthermore, it can undergo a chemical transformation through the years, that when evacuated it can combine with oil to make a rather dangerous explosive.

testing 230-volt equipment

You can test a 208/230-volt circuit in the same way as you would an ordinary 115-volt circuit. If you use a 115-volt bulb in the tester, however, it will burn out. A special 230-volt bulb is required. With it, you can insert the probes in a three-prong receptacle and find that, between one of the live probes and the ground, the light will light dimly, and, between the two live probes it will burn brightly.

A word of caution: *Before attempting to test or repair higher-voltage equipment, you would do well to secure the advice of someone who has had experience in such matters unless you are completely sure of what you are doing. Working on equipment of this sort presents the possibility of serious, even fatal, injury.*

You might be using a schematic which might be a bit confusing. Figure 8-6A is such a schematic and Fig. 8-6B shows the actual appearance of the parts and connections indicated by the schematic.

There are many ways of connecting motors with and without capacitors and relays, but in one particular situation, when you trace the current through the switches and protectors, you will find the motor is connected as shown in Fig. 8-7. Because there are three wires that are connected to the motor, it follows that you can use a three-wire test cord and bypass everything that you are directly concerned with.

"phases"

There is considerable confusion when reference is made to motor phasing. Just what does "phase" mean? The phases originate in the power company's generator. For explanatory purposes, we may assume that the generator stator (outside or stationary part) is divided into three segments, each with a wire leading from it. The rotor makes a revolution in one time unit. Suppose that there is an "X" on one of the two rotor segments, which is the one that rotates to produce the power you are using. Let us label the stator segments A, B, and C(see Fig. 8-8).

Each time the rotor segment marked "X" passes one of these stator segments, it sends out a cycle of a-c current. (The electron flow goes out, reverses itself, and then reverses again.) Now, because the rotor segment

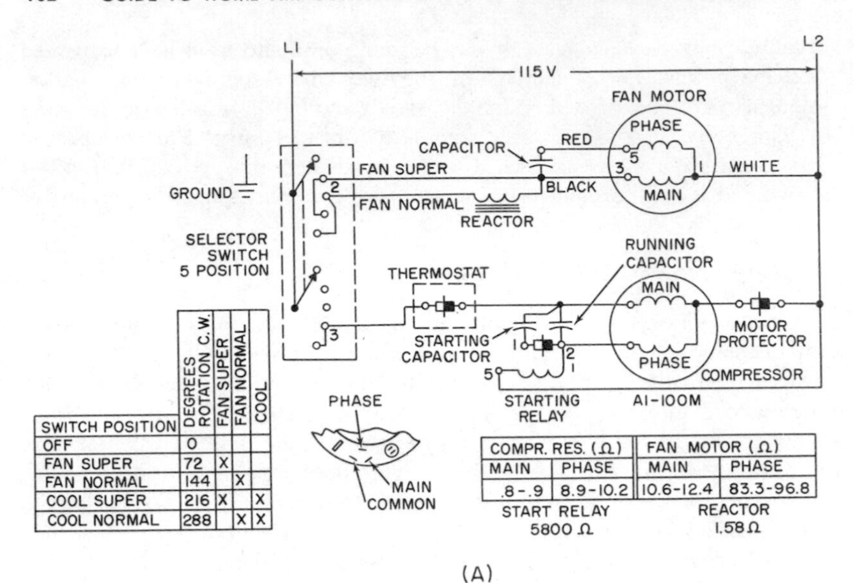

SWITCH POSITION	DEGREES ROTATION C.W.	FAN SUPER	FAN NORMAL	COOL
OFF	0			
FAN SUPER	72	X		
FAN NORMAL	144		X	
COOL SUPER	216	X		X
COOL NORMAL	288		X	X

COMPR. RES. (Ω)		FAN MOTOR (Ω)	
MAIN	PHASE	MAIN	PHASE
.8-.9	8.9-10.2	10.6-12.4	83.3-96.8

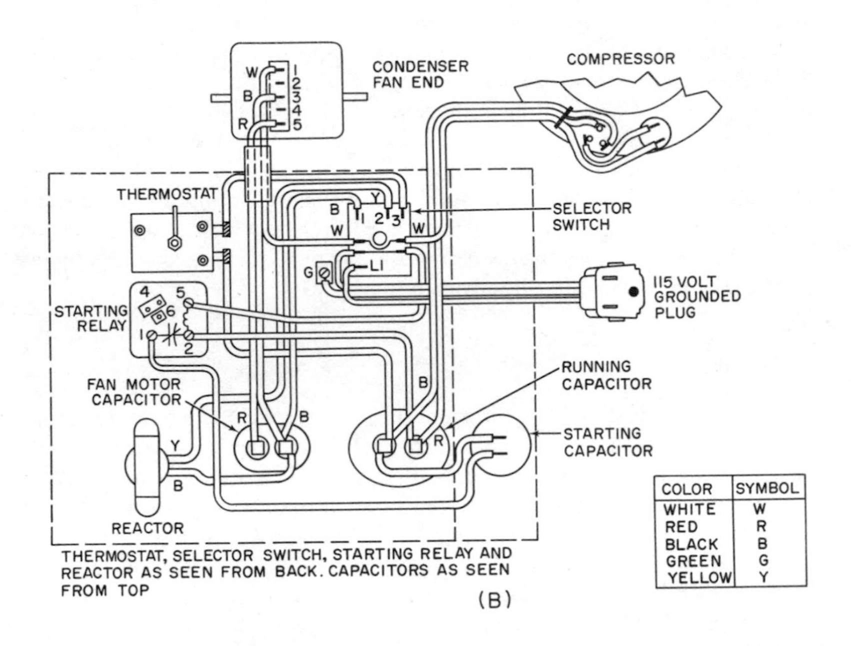

COLOR	SYMBOL
WHITE	W
RED	R
BLACK	B
GREEN	G
YELLOW	Y

Fig. 8-6. (A) A 115-volt circuit. (B) Component assembly of 115-volt circuit. (Courtesy of Frigidaire.)

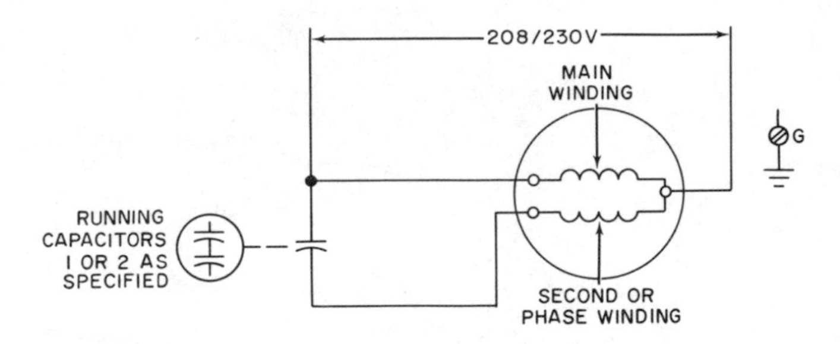

Fig. 8-7. One method of connecting 208/230-volt motor.

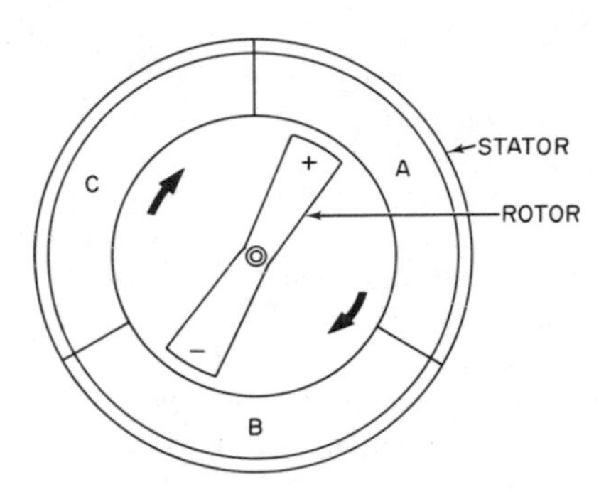

Fig. 8-8. A typical generator.

has to go past A to get to B, it follows that the B cycle is going to start a third of a unit of time later than the A cycle, and the C cycle is going to start two thirds of a unit of time later than the A cycle. "Staggering" the cycles in this fashion is called "phasing." The time unit referred to would be 1/60th of a second for 60-cycle current. Thus, the wire from A would begin a cycle, and 1/3 of 1/60th of a second later the wire from B would begin a cycle, and so on. In Fig. 8-9 we see a representation of the current flow. Each phase is represented by a line that curves and returns to its axis according to the reversal of electron flow.

208/230-volt systems

Whether 208 or 230 volts is delivered to your home is governed by the way the leads on the distribution transformer are connected. There are several different ways of connecting them. With one method, you get

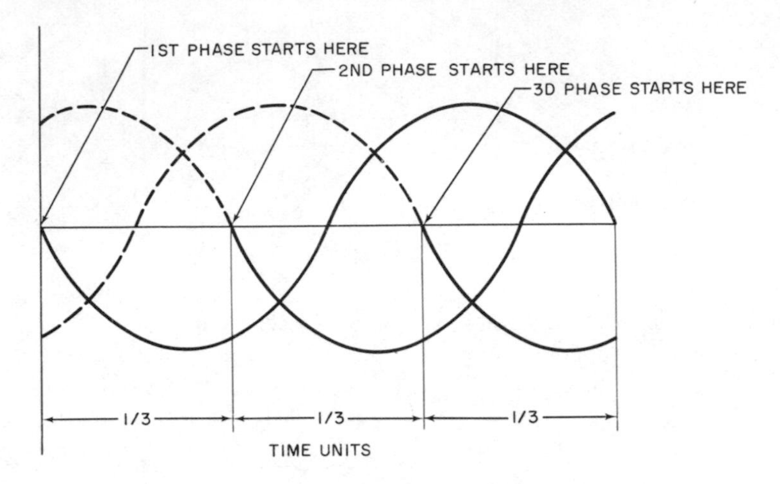

Fig. 8-9. Sine wave representation of three-phase current flow.

115 volts on an energized leg, and, using two of them, you get 230 volts. With another method, using two 120-volt legs, you get something less than twice 120. In this case it is 208 volts.

Figure 8-10 shows the possible connections on a four-wire, three-phase service. Now, one of these energized wires or legs can not do any useful work unless there is a circuit back to the generator (through the transformers). So, when one leg is used, the current has to flow through the so-called ground wire, and you get 115 volts.

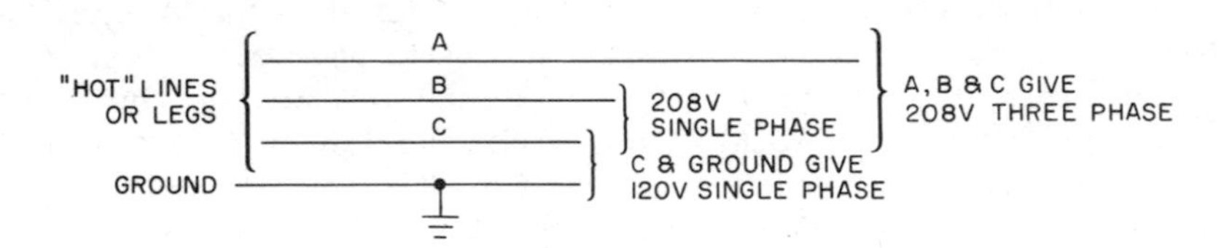

Fig. 8-10. 208-volt, four-wire, three-phase service.

In a 208/230-volt system, the two legs form the circuit, and the ground wire serves mostly as a way for voltage surges (such as lightning) to be carried away without damaging the equipment inside a building. On an air conditioner, the ground wire is attached to the chassis, and, somewhere on the other side of the receptacle, it would be connected to, possibly, a water pipe.

An ordinary three-wire system is called "single phase" because you can make a circuit with either phase and the ground wire. There also is a three-wire system in which all three of the wires are "live" and are connected to the motor. These "three-phase" motors are used on heavy equipment, with which we are not concerned. Having two different windings in a motor stator is called a "split phase" arrangement.

ohmmeter test for continuity

Perhaps the easiest way to check the continuity of a motor is with an ohmmeter like that shown in Fig. 8-11 or a volt-ohmmeter like

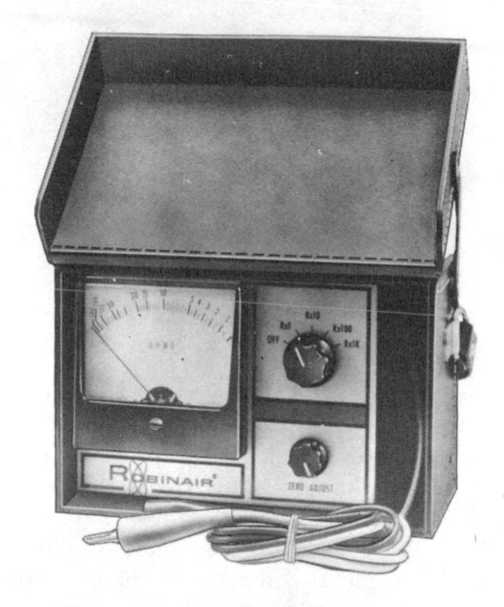

Fig. 8-11. An ohmmeter. (Courtesy of Robinair.)

that shown in Fig. 8-12. There should *not* be a circuit in a split-phase motor between the casing and a terminal. If there is, it means that an internal wire is rubbing on the casing and the insulation has worn off the wire, that is, it is shorted.

As for the windings, each one has a specified resistance in ohms. A handbook devoted entirely to motors might have a chart that would state, or give a close indication, of the resistances in ohms (Ω) of the particular motor that you were testing.

An ohmmeter test is made with the terminal wires removed. The meter has a little battery in it that sends out a steady stream of electrons (a direct current). If you run the current through a coil of wire, the dial will give a resistance reading in ohms. Direct-current resistances in series add up directly. (Alternating current is another matter.)

Figure 8-13 is a schematic of the windings of a motor with some resistance values assigned to them and the ground or common wire. (You would not actually have the values given; they have been chosen for explanation purposes). Now, with the probes on black and red,

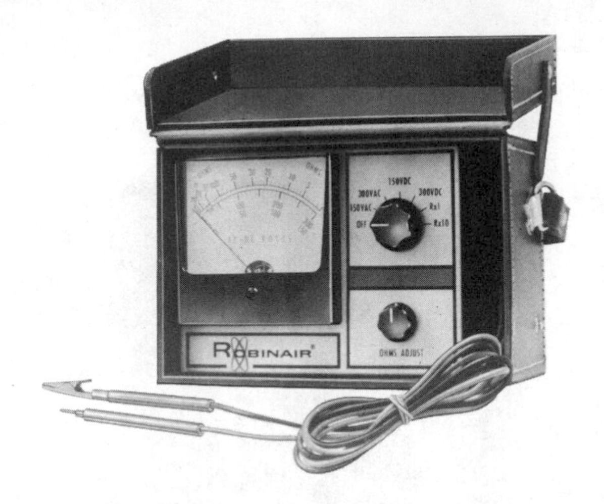

Fig. 8-12. A volt-ohmmeter. (Courtesy of Robinair.)

the reading should be 2 plus 3, or 5. By the same token, black and white are 3, and red and white are 4. If the windings were shorted so that they were, in effect, connected together, the reading would be something else, and, if the wire were broken, there would be no reading.

We have said that a motor-compressor can be started with a test cord. That is true, but some motors in sealed units have four terminals. In this case, the fourth one leads to a relay solenoid, which is only a different way of starting the motor. The fourth wire is on a third, or auxilliary winding. When the motor approaches rated speed, the generator effect spoken of previously (the so-called counter electromotive force) will energize the solenoid so that its plunger will raise. The plunger is attached to a disconnecting device, and that opens the contacts.

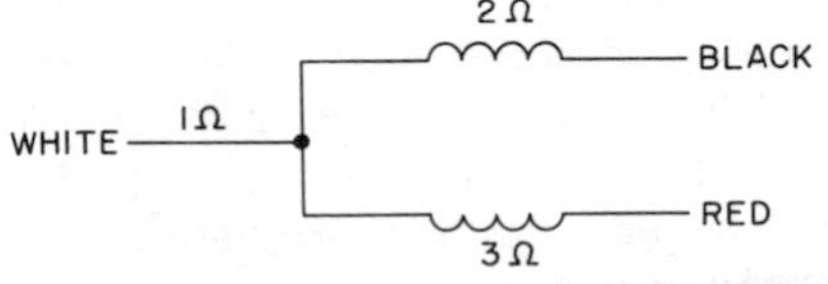

Fig. 8-13. Typical motor windings.

9—Freezing With Fire— The Gas Refrigerator

"absorption"

A book on cooling equipment would not be complete without devoting some space to so-called absorption equipment. Absorption refers to a chemical process that uses heat to produce cold! Absorption is the basis of the "gas" refrigerator. All of the domestic gases have been used for fuel as well as the electro-resistance coil, and kerosene-fired units have been used in remote places where electricity was not available.

The chemistry was worked out in a laboratory in the 18th century, and it was applied quickly. Coal-fired units were made that worked after a fashion. Then, in 1857, Pierre Carre patented an improved design, and others proceeded to secure patents by the dozen, on both sides of the ocean.

The chemistry has not changed much through the years; the constant redesigning was to secure better utilization of the fuel and more dependable operation. How does it work? Here is what the Morphy-Richards Company has to say about their Astral refrigerators (see Fig. 9-1):

"The sealed unit contains water-ammonia and hydrogen under pressure, and cannot be repaired in the field. The heat generated in the boiler causes ammonia vapor and water vapor to be expelled from the solution and rise into the rectifier. The water vapor is condensed in the rectifier and runs back into the boiler. The ammonia vapor passes through into the condenser tubes where it is liquified by air-cooling. The liquified

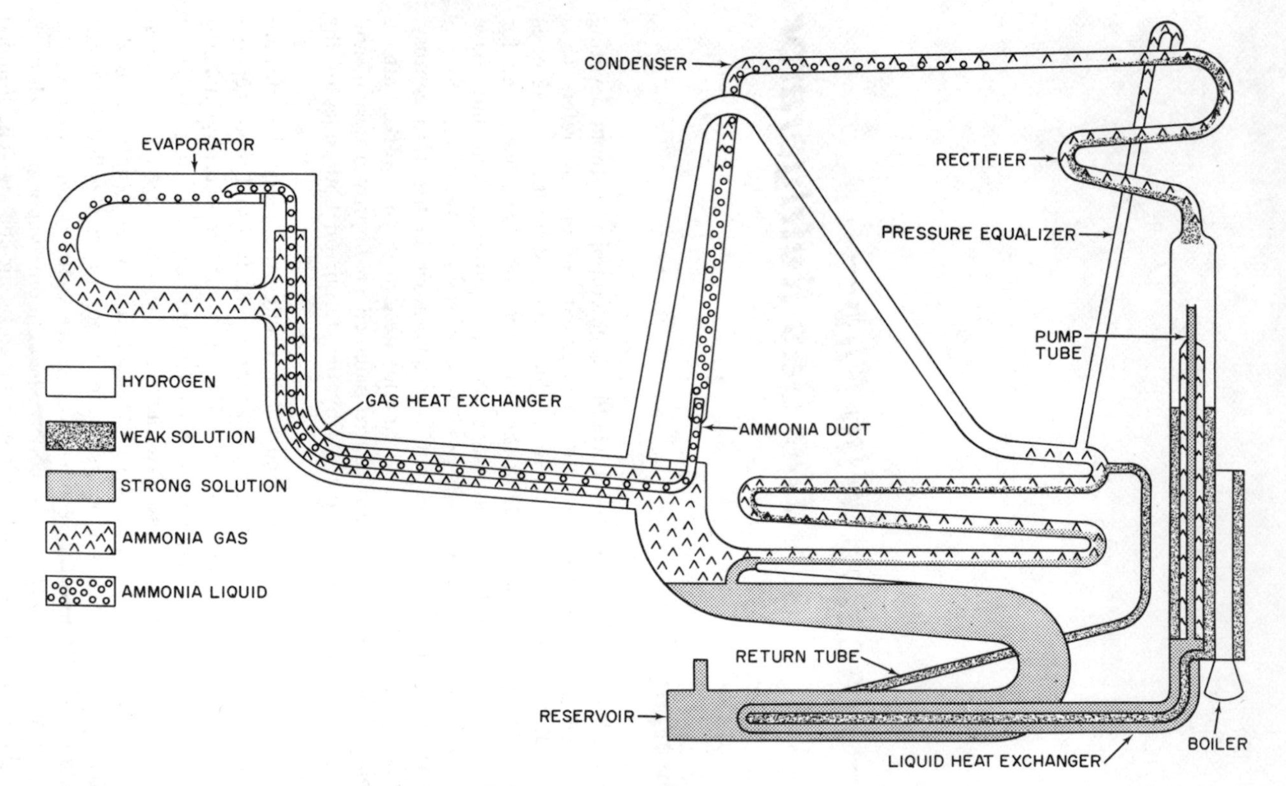

Fig. 9-1. Absorption type cooling unit. (Courtesy of Morphy Richards, Inc.)

ammonia is gravity fed through the ammonia duct to the top of the evaporator.

"Precooled hydrogen is introduced into the evaporator by way of the gas heat exchanger. In the evaporator the partial pressure of the ammonia is considerably reduced due to the partial pressure of the hydrogen (Dalton's Law of partial pressure). This results in the evaporation of the ammonia and consequent absorption of heat through the walls of the evaporator. A suitable cover is fitted over the evaporator to allow the maximum possible absorption to take place.

"The heavy mixture of hydrogen and ammonia vapor passes down through the gas heat exchanger into the lower end of the absorber where it meets a weak solution of ammonia flowing down through same. Due to the ammonia's great affinity for water, practically all of the ammonia vapor is absorbed and carried down to the reservoir, and the relatively pure hydrogen rises through the absorber and re-enters the gas heat exchanger, so completing the cycle.

"The strong ammonia liquid from the absorber passes down through the reservoir to the bottom of the boiler by way of the liquid heat exchanger where it is preheated and then passes to a small bore tube, the pump tube. The heat supplied to the boiler causes bubbles of ammonia gas to be formed, resulting in drops of liquid being forced out of the top of the tube into the outer boiler compartment (thermal-lift pump).

"Ammonia is again driven off into the condenser, and the weak solution sinks to the bottom of the boiler, into the liquid heat exchanger where it is cooled, and then passes up through the return tube into the absorber, which is arranged at a lower level than the pump to allow gravity feed to take place."

importance of "leveling"

The parts of the unit have been located so that a certain specified angle (from horizontal) will be obtained in the various tubes. If the refrigerator itself is not level, the vapor or liquid cannot flow effectively, and its efficiency will be poor; indeed; it may not freeze at all.

These gas-sealed units have the reputation of being exceedingly well built. After many years, however, they sometimes seem to have discharged, although it is likely that the chemicals separated. Or, possibly, an impurity caused a reaction so that the chemical balance became upset. In any event, that condition can be remedied by turning the unit upside down. How's that again? Rotate it 180 degrees so that the top rests on the floor, and just leave it alone for a few days. Many units have been successfully restored to operation this way.

testing the cold control

Here, as with the mechanical refrigerator, you can suspect the cold control in case of incorrect cabinet temperature. Let us look at an electric-coil type of absorption unit. The input would be 115 volts ac if it plugs into a wall receptacle, or it could be 115 or less d-c volts. It does not matter which or what voltage as long as you have the right bulb for your two-wire tester. The thermostat would make-or-break the feed to the resistance heater, and you can use the two-wire tester to determine if the heater is being energized simply by placing the two probes on the two terminals. If it is energized, and it does not give off any heat (which should be obvious if you feel it), the resistance wire must be broken. If it is not energized, go to the cold control and touch the terminals on it. If the bulb still does not light, there must be a broken wire some place. When there is a broken wire inside the insulation, the the only way to isolate it is to check out the circuit. To do that (assuming the bulb socket has short leads) you must put an extension of, say, two feet on one of the leads. This can be done with another piece of wire with a clip on each end. Clip the loose end to one of the input terminals, and touch the probe to various points around the circuit. When you get to the point between where the bulb lights and where it does not, that is where the trouble is.

If you think that the switch contacts are not closing when they should, put the probes back on the resistance heater, and turn the cold control dial a little. If turning it, say, a quarter of an inch will energize the heater, the control is simply calibrated wrongly and would serve well enough on a slightly colder setting. If you have to give it, say, half a turn beyond the normal setting, it would appear to have the classic symptoms of a discharged bellows. By the same token, over cooling might be the result of a faulty switch.

Checking the switch cycle (i.e., the percentage of time that it runs) might appear to pose a problem because the unit would be silent in operation; but there is a way to do it. If the voltage is 115, you can use an electric clock! The clock wires have to be connected so that it only runs when the heater is energized, of course. If there is any question in your mind, just put the tester probes on the heater, turn the dial back and forth, and notice if the clock runs when the bulb lights and is stopped when the bulb is out.

heat input

Absorption refrigerators run on a rather long cabinet-temperature cycle. That is, it takes quite a while after the heat goes on before the freezer temperature changes radically. This type of unit is not like the mechanical type in which, after the compressor starts, you can touch the evaporator and (with certain models) feel the metal cool almost within seconds. An absorption unit is designed to operate with a specified heat input. The heaters come in various wattage ratings for the different cabinet sizes and voltages encountered. If the unit was supposed to run say, 40% of the time, and, because the heater is of too small a rating, it ran 75% of the time, it follows that the freezer temperature would drop slower than is normal and the cabinet temperature would rise higher than is normal as a consequence. In an extreme case, this situation could result in a tendency to defrost, which might be mistakened as a symptom of a partly discharged switch bellows.

Although some variation in cabinet temperature and upsetting of the convection pattern might result, attention to correct heat input is mostly a matter of economy of operation. The right amount of heat is needed to have the chemicals flow in the tubes in an orderly fashion. If the heater is supposed to run on 115 volts and only 100 volts is being delivered, there would be a difference in the monthly power consumption. This is a case in which attention to proper wiring will pay dividends.

The proper wattage is supposed to be indicated on a small plate attached to the unit. The actual heat output in watts can be computed. All that is required is a voltmeter and ammeter. With the unit wired so that the current in one of the leads will flow through the ammeter, the reading in amperes multiplied by the known voltage will give an answer in watts. If it is more than 10% off, the unit needs attention.

Excessive operation with too cold a cabinet could be due to electrical leakage (short) between the cold-control leads, or the capillary-tube bulb might be loose and the heat transfer reduced by corrosion or ice. The bulb is supposed to be tight against the evaporator. The bulb would not be able to sense that the freezer was cold enough for the unit to shut off and could not indicate this condition because of poor contact.

gas burners

Most absorption units run on some form of domestic gas. There are four broad classifications: manufactured (city) gas, natural gas, propane, and butane. Various burners are capable of using each class. The burners are somewhat similar in appearance, but you can not substitute one for another because the opening that controls the gas flow is different in each. This "orifice" is in a small brass screw-in fitting, often called a "spud." In each class of burner, minor variations (mixtures) in the gas supply can be compensated for by changing spuds. The amount of air required for combustion differs with each class, and the manner of mixing the air with the gas before it burns (primary entrainment) varies also. Thus, you will find burners with a different internal arrangement with respect to the ports (air passages). People have made unauthorized substitutions, and, even though the burner could be adjusted so the flame looked all right, it produced carbon that filled the flues and cut down the draft, which resulted in even poorer combustion. With the wrong burner, some carbon monoxide would probably be produced (which you would not smell) and something else, which, impolitely stated, stinks to high heaven. It is a biting odor that, once smelled, will never be forgotten.

Gas-refrigerator controls are about the same as those on a gas water heater. There are some variations according to make and model, but, in general, to start, you push a button to allow a small amount of gas to flow from the "pilot" tube. This gas flow is ignited with a match. The flame will strike a "heat conductor" that will carry the heat to a bimetallic disc that, when heated, will warp, pulling its attached plunger away from a valve seat and opening the main gas supply to the burner. With the main valve open, and the thermostat not calling for operation of the unit, the gas will flow through a bypass and support a small flame (the minimum flame) in the burner. The flame at the "pilot" tube will ignite the burner, and, at this point, the button can be released. The burner flame will maintain enough heat on the heat conductor so that it will keep the valve open.

When the thermostat bellows expands, it has the effect of opening the gas port for a larger, or maximum, flame. The two flame sizes, (usually called low and high) are regulated by screws. How much flame you will have with a particular screw setting is relative to the pressure of the gas in the supply line. For efficiency and economy of operation, the flame sizes are supposed to be held within very close limits.

the manometer

The flame size is not a matter of dimensions, it is what you get with any given pressure. To make pressure adjustments, you need a manometer like the one shown in Fig. 9-2. This manometer is just a U-shaped length of glass tubing mounted on a board that is marked in inches so that you can read the difference between the tops of the two water columns. One glass-tube end has a rubber tube on it and the other is open to the air. The rubber tube connects to an outlet on the burner or controls. If none is readily found, it is because the manufacturer has supplied a screw-in fitting for this purpose.

To take a reading, put water in the tube until it reaches the zero mark (or thereabout). With the gas outlet open, the gas will flow through the tube and push one column down and the other up. If one reads 3 inches above zero and the other 1½ inches below zero, you have 4½ inches of gas pressure. The pressure in the main intake pipe is supposed to be 11 inches for propane, for example, and should hold very close to that

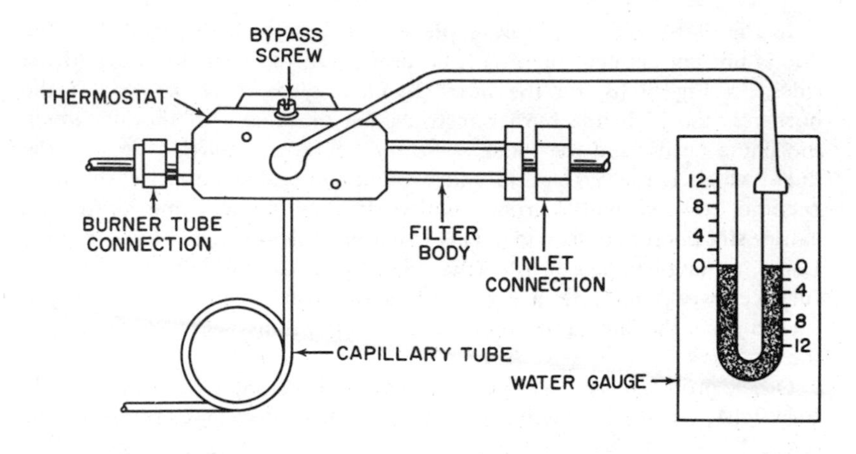

Fig. 9-2. Manometer or water gauge connected to thermostat. (Courtesy of Morphy Richards, Inc.)

as the high flame goes on and off. Readings for other gases vary so much that it is not even worthwhile to give the general order of magnitude. You will have to get the correct reading from the dealer or manufacturer.

The main service problems on absorption units derive from an accumulation of dirt (carbon) or the burner going out. The two are somewhat related. If the heat conductor gets too much carbon on it, the carbon will cut down the heat transfer and allow the conductor to cool off, thereby closing the bimetallic disc and shutting off the gas supply.

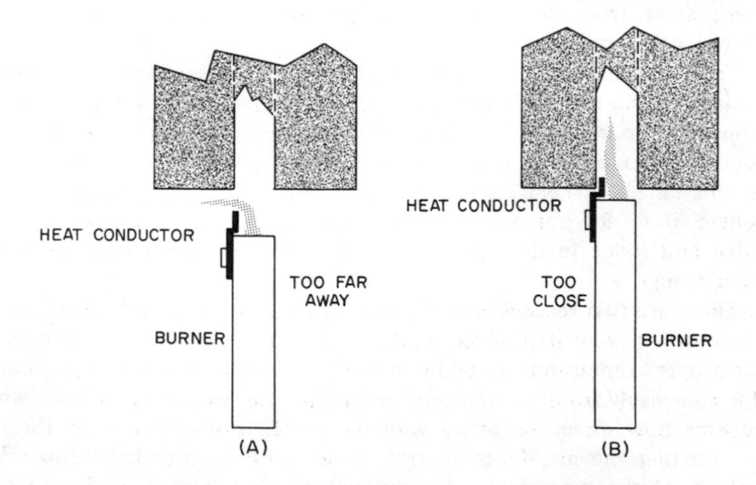

Fig. 9-3. Burner placement for vertical flue opening.

In Fig. 9-3A, we see a burner placed too low, with the result that the flue is not hot enough, there is little draft, and the heat dissipates to the sides. In Fig. 9-3B, on the other hand, we see what happens if the burner is too high; the burner restricts the movement of the air, which the flame needs, and the flame is "dirty." Smoke is carried up into the flues, which is the passage in and around the sealed unit, and the flues become covered with carbon, which further reduces the draft and causes still poorer combustion. This situation calls for a chimney cleaning. Using a long-handled brush (the kind used on the old coal-burning furnaces works fine) or a cord with a rag secured in the middle, you should clean the flues from top to bottom. (It will be necessary to remove the burner.)

Other problems are encountered with a horizontal flue opening. In Fig. 9-4A, we see a situation in which the heat does not enter the flue.

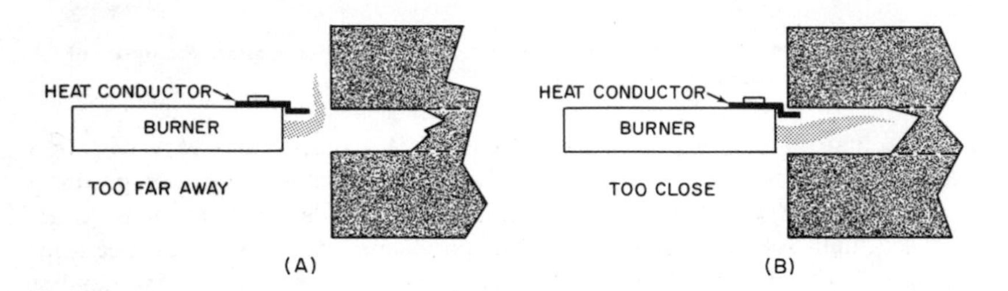

Fig. 9-4. Burner placement for horizontal flue opening.

In Fig. 9-4B, the burner is too near the flue, and the draft carries the flame away from the conductor, which, of course, could result in the burner going out.

Then again, it might happen that the heat conductor was adjusted during the summer and the much lower temperature of the air drawn from the floor during the winter was sufficient to reduce the heat-conductor temperature below the point required to keep the valve open. In this case, the heat conductor can be bent for best performance. Of course, it is not impossible for a speck of rust to be carried into the valve and lodge in the bypass opening. This situation can be remedied by cleaning.

There are two reasons why the flame might be dirty and smoky, or, in other words, why it produces carbon. One reason is lack of enough air for proper combustion. In addition to the situation referred to previously, this can result from an improper setting of the burner air intake, which governs how much air mixes with the gas before it burns. In Fig. 9-5 we see that the air shutter barrel screws onto a threaded fitting. With burners of the type shown, the amount of air mixed in is governed by

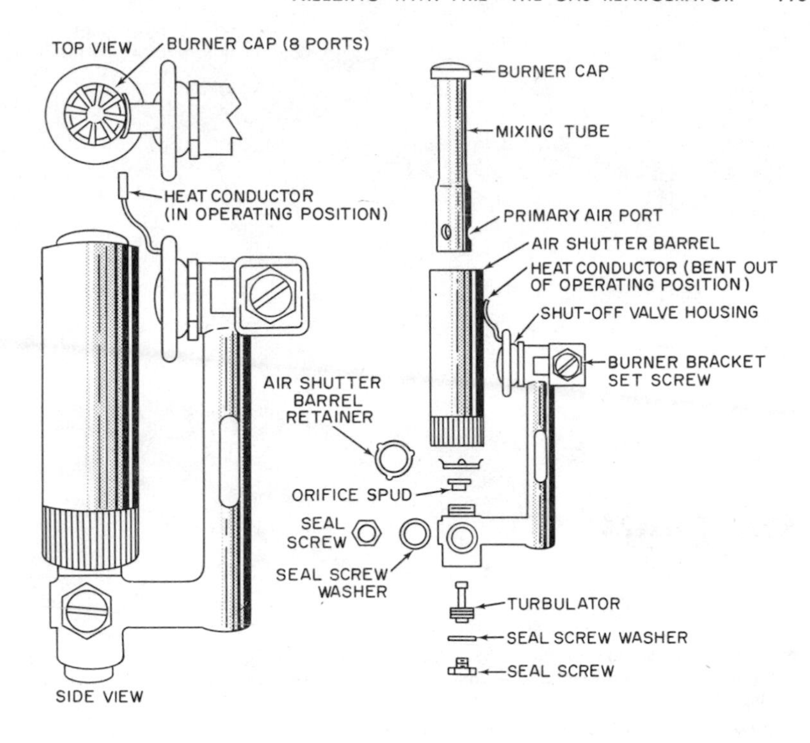

Fig. 9-5. Gas burner component assembly.

turning the barrel. When the burner is turned all the way to the bottom, the flame (for lack of enough air) will be yellow and smoky. Rotating the barrel, so that it goes up to the top of the shaft, on the other hand, will give the flme too much air and it will be noisy and "fluttery." The correct setting is about half-way between these two extremes. The other reason for a carbonizing condition is that the flame is hitting the side of the flue. To correct this, the burner can be moved to center the flame in the flue opening.

Glossary

Bulb: The end portion of the small tube on a thermostatic switch. In the early days the tube was enlarged at the end; most modern switches have it coiled or merely looped.

Capillary (Tube): A tube of very small diameter. The term is applied to the tube on a thermostatic switch, the spiral passage in a restrictor, the tubes from the restrictor to the evaporator (where used), and could refer to the restrictor and liquid line together.

Carbon: An element, black and sooty, formed by smoke from a dirty flame.

Carbonized (Oil): A lubricant that has become sticky.

Charge: The specified amount of refrigerant in a system. Putting in the refrigerant is called "charging," and when it has leaked out or removed, the system is "discharged."

Circuit: The path taken by a drop of Freon as it makes its way through the various parts of the system and returns to the point from which it started. An electrical circuit is a path through which a current is flowing.

Coil: A length of wire or tubing that is wound in a circular fashion like thread on a spool. In an air-cooled condenser or evaporator, the coil is the tubing, which is arranged in a number of loops.

Cold Control: Same as thermostatic switch.

Components: In a refrigerator, the sealed unit, cold control, and relay. In a sealed unit, the motor-compressor, evaporator, and condenser.

Compressing: Causing gas or air to occupy a smaller space by exertion of pressure.

Condensing: Causing a vapor to become a liquid by application of pressure or lowering its temperature.

Condensate: The water that drips from an air conditioner's cooling coil.

Cycle: The period of time required for a unit to start, run, shut off, stand idle, and start again.

Energize: To apply power to something electrical, such as a solenoid coil. If it has no current running through it because it is shut off, or because of a broken wire, etc., it is said to be de-energized.

Evacuate: Use a vacuum pump to remove the Freon and air, if any, from the interior of a refrigerating unit.

Evaporator: The part of the system in which the liquid Freon "boils" and turns to a vapor.

Fins: Pieces of sheet metal secured to tubing to help it give off heat, or to cool the air, as the case may be.

Freon: The most common refrigerant.

Gas: A general term applied to refrigerants. In an absorption-type refrigerator, it would refer to the fuel that is burned to produce heat that produces cold.

High Side: The part of the system where a high pressure prevails, i.e., in the compressor, condenser, liquid line, and restrictor.

Liquid Line: The tube that carries liquid Freon from the condenser to the restrictor. It is much smaller than the suction tube that carries the Freon vapor back.

Low Side: The part of the system where low pressure prevails, i.e., in the evaporator and the suction line leading back to the compressor.

Pass: One loop of tubing.

Pressure, Back: The pressure found on the low side.

Pressure, Head: The pressure found on the high side.

Refrigerant: The fluid (liquid or vapor) that is used to create cold by absorbing heat at a low temperature and pressure in one place and giving off the heat at a higher temperature and pressure at another place.

Relay: A device to help the motor to start.

Resistance Heater: An electric heater that has one or more wires that become red hot when energized because of the resistance to the flow of current by the wires.

Restrictor: A device that keeps the liquid Freon from running into the evaporator too fast by making it go through a tube of small diameter.

Sublimation: Process by which a solid is transformed to the gaseous state without passing through the liquid state.

Suction Line: See Liquid Line.

System: The interior passages through which the Freon passes as it makes its way around the circuit.

Thermally Actuated (Switch): A device that will open (or close) contacts when the temperature changes. Sometimes called a thermo-switch.

Unit, Open: The obsolete construction in which the component parts could be removed by the use of wrenches.

Unit, Sealed: The modern construction in which the motor and compressor are in a welded shell and all tubing joints are soldered.

Vapor: Any gaseous substance. Any liquid will "boil away," or vaporize, at some temperature or other. Some solids, e.g., "dry ice" will sublimate, i.e., go from a solid to a vapor without becoming a liquid in between.

Index